Thisbe K. Lindhorst

Essentials of Carbohydrate Chemistry and Biochemistry

Thisbe K. Lindhorst

Essentials of Carbohydrate Chemistry and Biochemistry

Second,
revised and updated edition

Prof. Dr. Thisbe K. Lindhorst
Institute of Organic Chemistry
Christiana Albertina University Kiel
Otto-Hahn-Platz 3
24098 Kiel
Germany

Cover picture: Olaf Krohn, sugar cane harvest in Mauritius.

Library of Congress Card No.: applied for

A catalogue record for this book is available from the British Library.

Bibliographic information published by Die Deutsche Bibliothek
Die Deutsche Bibliothek lists this publication in the Deutsche Nationalbibliografie; detailed bibliographic data is available in the Internet at http://dnb.ddb.de

Printed in the Federal Republic of Germany.
Printed on acid-free paper.
Printing: Strauss Offsetdruck GmbH, Mörlenbach. Bookbinding: J. Schäffer GmbH & Co. KG, Grünstadt.
ISBN 3-527-30664-1

*Dedicated to Moritz and Justus
with love*

Preface

While this book has been written mainly for the non–carbohydrate or aspiring carbohydrate chemist, the reader will require a basic knowledge of organic chemistry. The book covers the main topics, but not all essentials of the field, emphasizing the more modern and frequently used methods. It is hoped that as well as providing an introduction to carbohydrate chemistry, this book will also provide incentive for further studies and reading and may be even of value to the experienced carbohydrate chemist as a useful reference text.

The structure of the book's contents has been kept simple as it was the intention to describe the structure and reactions of carbohydrates in relation to their biological activity. In this aspect which makes carbohydrate chemistry the intriguing field it is today, chapters 6 and 7 therefore pay tribute to the interdisciplinary field of glycobiology.

Many practical aspects including a number of experimental procedures have been included in the book and these should provide the basis for the reader to begin practical work in the laboratory. Guidance in NMR analysis of carbohydrate derivatives has also been provided. A number of primary literature citations are included, although the vast majority of the carbohydrate literature and most of the names of researchers contributing to the field could not be reflected. The final chapter however, introduces the most notable journals, book series and review articles in the field of carbohydrate chemistry which will open the door to the world of current carbohydrate research and those engaged in it.

Despite all the facilities available, it has always been difficult to start as a newcomer in carbohydrate chemistry and one cannot do better than to have a good teacher (and a good book!) at one's side. I am grateful to Professor Joachim Thiem who gave me counsel and guidance in carbohydrate research, which eventually led me to work in the neighborhood of Professor Hans Paulsen's laboratory at the University of Hamburg and furthermore, to write the 1st edition of this book in the almost historical office of Emeritus Professor Kurt Heyns, the 'scientific grandchild' of Emil Fischer. I would also like to thank my lecturers at Wiley-VCH for their patience and assistance during the long evolutionary process of the preparation of this book. I am grateful to my co–workers and students who have contributed to the book in various ways and who have helped to identify the mistakes in the 1st edition.

The 2nd edition of *Essentials of Carbohydrate Chemistry and Biochemistry* has been prepared in Kiel at Christiana Albertina University. It was intended to keep the handy formate of the book and a manageable list of content. Nevertheless the text has been revised, extended and updated here and there and some recent aspects of carbohydrate chemistry have been included. I wish, that also this 2nd edition will be a valuable book for chemistry and biochemistry students and will allow joyful reading and studies!

Kiel, Summer 2002   Thisbe K. Lindhorst

Contents

Preface

Abbreviations

Abbreviations

AcCl	acetyl chloride
Ac$_2$O	acetic anhydride
AgOTf	silver triflate
ATP	adenosine triphosphate
Boc	*tert*–butoxycarbonyl
Boc$_2$O	di–*tert*–butyldicarbonate
Bn	benzyl
BnCl	benzyl chloride
BzCl	benzoyl chloride
CAN	cerammonium nitrate
CD	cyclodextrin
CDA	cyclohexane–1,2–diacetal
CMP	cytosine monophosphate
COSY	corellated spectroscopy
CRD	carbohydrate recognition domain
DAST	diethylaminosulfur trifluoride
DBE	dibromomethyl methylether
DBU	diazabicycloundecen
DCC	dicyclohexylcarbodiimide
DDQ	2,3–dichloro–5,6–dicyano–1,4–benzoquinone
DEAD	diethylazodicarboxylate
DHP	dihydropyran
DISPOKE	dispiroketal
DMAP	dimethylaminopyridine
DMDO	3,3–dimethyldioxirane
DMF	dimethylformamide
DMSO	dimethyl sulfoxide
DMTST	dimethylmethylthiosulfonium trifluoromethanesulfonate
Dts	dithiasuccinoyl
ER	endoplasmic reticulum
ESI	electrospray ionization
EtOH	ethanol
Fmoc	fluoren–9–yl–methoxycarbonyl
GPC	gel permeation chromatography
GPI	glycosylphosphatidylinositols
HMDS	hexamethyldisilazan
HOAc	acetic acid
HPLC	high pressure liquid chromatography
IDCP	iodonium dicollidine perchlorate
i–Pr	isopropyl
IUPAC	International Union of Pure and Applied Chemistry

Kdo	3–deoxy–α–D–manno–oct–2–ulopyranosonic acid
MALDI	matrix–assisted laser desorption ionisation
MCPBA	*m*–cloroperbenzoic acid
4–Me–DTBP	2,6–di–*tert*–butyl–4–methylpyridine
MeOH	methanol
MeOTf	methyltriflate
Mesyl	methanesulfonyl
NMR	nuclear magnetic resonance
NBS	*N*–bromosuccinimide
NIS	*N*–iodosuccinimide
ORD	optical rotation dispersion
PAMAM	polyamidoamine
Ph	phenyl
Pfp	pentafluorophenyl
p–MBn	*p*–methoxybenzyl
PrOH	propanol
p–TsOH	*p*–toluenesulfonic acid
TASF	tris(dimethylamino) sulfur (trimethylsilyl)difluoride
TBABr	tetrabutylammonim bromide
TBDMS	*tert*–butyldimethylsilyl
TBDPS	*tert*–butyldiphenylsilyl
t–Bu	*tert*–butyl
TCP	tetrachlorophthalimido
TES	triethylsilyl
TESOTf	triethylsilyl trifluoromethanesulfonate
TFA	trifluoracetic acid
TfOH	trifluoromethanesulfonic acid
THF	tetrahydrofuran
THP	tetrahydropyranyl
TIPS	triisopropylsilyl
TIPDS	tetra–isopropyl–disiloxanylidene
TLC	thin layer chromatography
TMS	trimethylsilyl
TMSCl	trimethylsilyl chloride
TMSET	2–(trimethylsilyl)ethyl
TMSOTf	trimethylsilyl trifluoromethanesulfonate
Tosyl	*p*–toluenesulfonyl
Triflyl	trifluoromethanesulfonyl
UDP	uridine diphosphate

1   Introduction

Carbohydrates comprise the most abundant group of natural products. They are prime biological substances, of which billions of tons are produced every year by photosynthesis by plants and microorganisms. They form the major constituents of shells of insects, crabs, and lobsters, and the supporting tissue of plants but moreover, they are present as parts of basically all cell walls, spanning from the world of microbes to mammals.

The generic term 'carbohydrates' includes monosaccharides, oligosaccharides and polysaccharides as well as substances derived from monosaccharides by reduction of the carbonyl group (alditols), by oxidation of one or more terminal groups to form carboxylic acids, or by replacement of one or more hydroxyl groups by hydrogen (deoxy derivatives), an amino function, a thiol group, or similar heteroatomic groups. Carbohydrates exist in a large constitutional as well as stereochemical variety, as they are built up from monosaccharides of various kinds, forming branched or linear oligomers, as well as polysaccharides.

Carbohydrates possess a large number of functionalities, at least one carbonyl and several hyroxyl functions per monosaccharide, and often even carry further kinds of functional groups. In addition, they are compounds with several stereocenters and thus the carbohydrate group consists of a large number of stereoisomers. Synthetic carbohydrate chemistry, as a result of the structural complexity of carbohydrates, has to deal with two main problems, (i) the regio– and stereoselective formation of glycosidic linkages for the synthesis of oligosaccharides and (ii) the selective protection and deprotection of saccharide building blocks prior to and after the linkage step. This makes carbohydrate chemistry a difficult field for organic chemists.

The initial carbohydrate chemistry dealt with the structure of carbohydrates and solved basic questions of the stereochemistry problems connected with it. This was mainly due to Emil Fischer, who solved all these basic questions by the end of the nineteenth century. Later in the 1960s all main aspects of the roles played by carbohydrates in the storage and in the supply of energy in biochemical systems were known. Biosynthesis and biodegradation of carbohydrates became clear. The art of chemical transformation of monosaccharides as well as oligosaccharide synthesis was further developed, motivated by the isolation of biologically active compounds from microorganisms, such as antibiotics, which contained unusual saccharides.

However, an important next step in the development of carbohydrate research only followed thereafter. It was the discovery of the biological importance of glycoconjugates containing a great structural variety of complex oligosaccharides, which are used as ligands in biologically important molecular recognition processes. These findings have triggered a renaissance in carbohydrate chemistry, also supported by the improvement of analytical methods such as chromatographic techniques, NMR, and mass spectroscopy which allowed the ready purification of the normally non–crystalline compounds and the elucidation of their complicated structures.

Today, carbohydrate–based recognition phenomena form the basis of a modern research field, named glycobiology. It will require an interdisciplinary endeavor to unravel the diverse functions encoded by complex carbohydrate structures found in glycoconjugates and to manipulate them in a therapeutic way, where appropriate. The aim of this book is to give an overview of carbohydrate and glycoconjugate structures and the basic concepts of their synthesis, to enable the synthetic chemist to understand and, possibly, to contribute to glycobiological research.

2 Structure of saccharides

2.1 Structure of monosaccharides

Monosaccharides are the chemical units from which all members of the carbohydrates are built. The name 'sugar' is often used as a synonym for carbohydrates in general, while in everyday usage it means the table sugar sucrose. The simplest sugars are polyhydroxycarbonyl compounds with the general molecular formula $C_n(H_2O)_n$. Deduced from this formula (which does not hold for the majority of carbohydrate derivatives), carbohydrates have been named 'hydrates of carbon' as literally in the German word for carbohydrate, *'Kohlenhydrat'*.

Carbohydrates exist as aldoses, which are polyhydroxyaldehydes, and ketoses, being polyhydroxyketones. Their open–chain forms can be suitably represented by Fischer projections. Aldoses and ketoses can be of different chain length and are all derived from glyceraldehyde. A formal insertion of a stereogenic (HCOH)–group between the carbonyl carbon atom and the α–C–atom of glyceraldehyde leads to aldotetroses and further successive (HCOH)–group insertions between the carbonyl carbon atom and the adjacent stereogenic center to aldopentoses and aldohexoses. By an analogous sequence the group of ketoses branches out from 1,3–dihydroxyacetone, leading to tetruloses, pentuloses, and hexuloses.

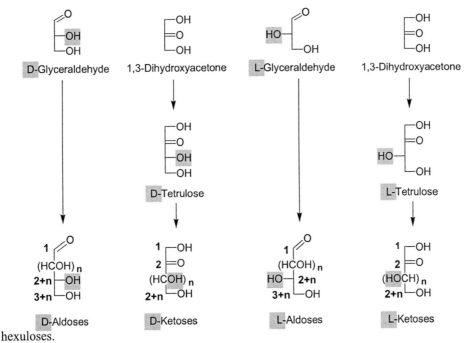

hexuloses.

The group of aldoses is derived from glyceraldehyde, ketoses can be built up from 1,3–dihydroxyacetone.

Thus a structural tree is created of which every member is characterized by a distinct sequence of stereogenic centers, each of which was attributed an individual name. However, each aldose and ketose exists as two enantiomers. To distinguish between the two, the prefixes 'D' and 'L', respectively, are used. D and L are relative designations for the configuration of the highest–numbered stereogenic center of a monosaccharide, in other words the one most distant from the carbonyl group, which is numbered 1 in aldoses and 2 in ketoses. This relative nomenclature is based on the configuration of the two enantiomeric glyceraldehyde molecules in the case of the aldose family and on the stereochemistry of tetrulose for ketoses. When the hydroxyl group at the stereogenic center of glyceraldehyde (and tetrulose, respectively) is positioned left in the Fischer projection, the molecule is named the L– (latin: *laevus* = left) enantiomer, when this hydroxyl is positioned right, its called the D– (latin: *dexter* = right) form. Certainly, D– and L–sugars could also be described using the (R)– and (S)–designations according to the Cahn–Ingold–Prelog rules. The historical D–, L–convention, however, is advantageous in carbohydrate nomenclature and has, therefore, been maintained. In addition to the D–, L–designations, carbohydrates are often specified by the sign of their optical rotation, which certainly bears no logical relationship to the D–, L–convention.

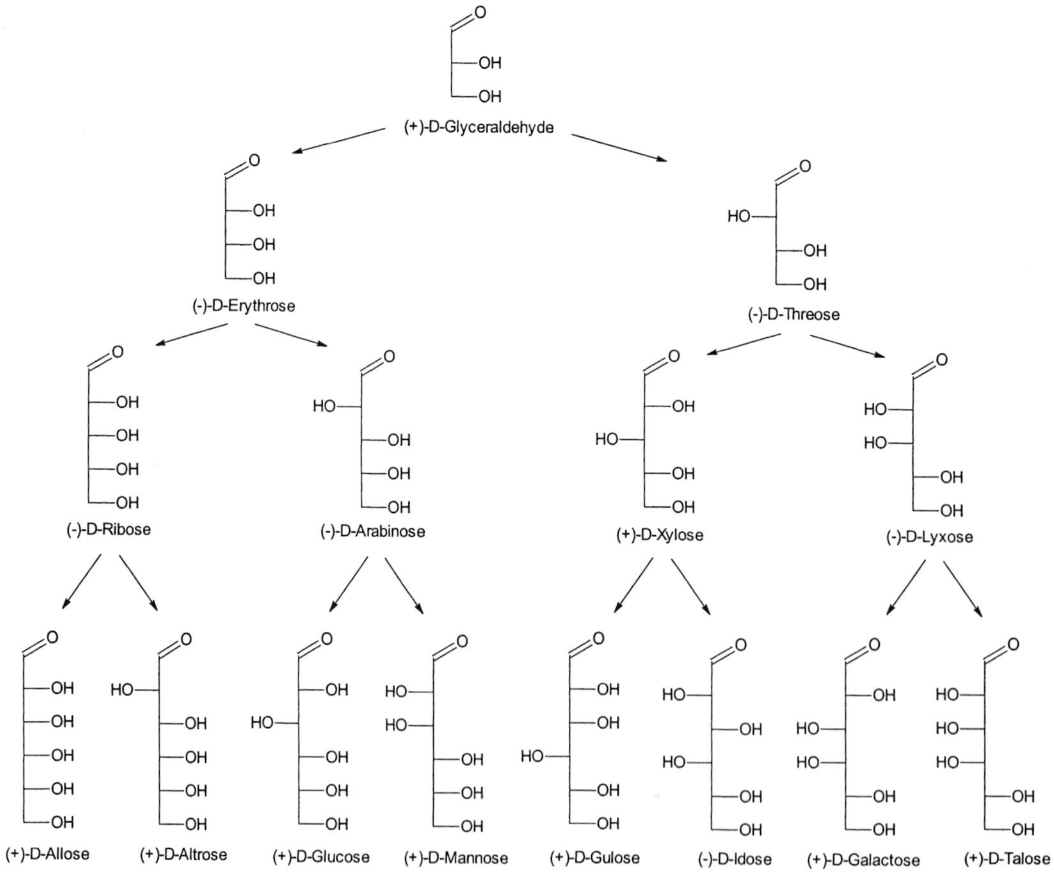

The acyclic forms of D–aldoses drawn as their Fischer projections.

In greater detail, monosaccharide structures are deduced from glyceraldehyde as follows: Glyceraldehyde has one stereogenic center and consequently occurs in two enantiomeric forms, which are (+)–D–glyceraldehyde with the (R)–configuration and (-)–L–glyceralde-hyde with (S)–configuration. Formal (HCOH)–group insertion between the carbonyl carbon atom and the adjacent stereogenic center leads to two diastereomers with the D–configuration emerging from D–glyceraldehyde and two others with the L–configuration from L–gly-ceraldehyde, which are enantiomeric to the respective former diastereomers. These four stereoisomers form the group of aldotetroses. Further chain elongation gives rise to eight stereoisomeric aldopentoses, four of them with D–, the other four with L–configuration. In the next step 16 stereoisomeric aldohexoses are created, eight D– and eight L–configured. Monosaccharides with more than six carbon atoms are relatively rare.

An analogous procedure leads to the family of ketoses starting with two enantiomeric tetruloses. Four pentulose stereoisomers are derived upon the first (HCOH)–group insertion and eight stereoisomeric hexuloses in the next step. As ketoses always carry one stereogenic center less than the aldoses of the same chain length, only half the number of stereoisomers arise in every series compared to the isosteric aldoses.

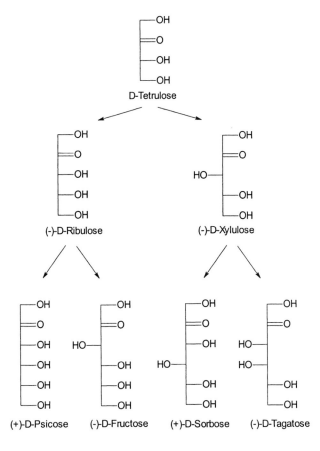

The acyclic forms of D–ketoses drawn as their Fischer projections.

Thus, an enormous stereochemical variety of aldose and ketose structures results, providing a fruitful area for the study of stereochemistry. D–Glucose, for example, the most abundant monosaccharide found in nature, contains six carbon atoms and four stereogenic centers in the open chain form. Consequently, it is just one stereoisomer from a group of 16 stereoisomeric aldohexoses which all have the molecular formula $C_6H_{12}O_6$. Most of the widely distributed monosaccharides belong to this group of hexoses. In addition, there exist eight stereoisomeric hexuloses (ketoses with six carbon atoms) which are constitutional isomers to the aldohexoses. The 16 stereoisomeric aldohexoses consist of eight enantiomeric pairs, one enantiomer being the L–form, its mirror image comprising the D–enantiomer in each case. Aldohexoses, which are not enantiomeric to each other are called diastereomeric. In carbohydrate chemistry the term 'epimeric' is also frequently used. Epimers are diastereomeric compounds which differ only in their stereochemistry at the stereogenic carbon atoms adjacent to the anomeric centres (the carbonyl carbon atom). Isomers which have a different configuration at one other carbon atom have been named 3–, 4–, or 5–epimers, respectively.

Fischer's elucidation of the basic stereochemical interrelationships of monosaccharides

The structures of glucose and its isomers were elucidated at a time when no NMR spectroscopy and modern separatory techniques were yet known. It was Emil Fischer who discovered the interrelationships of the large family of monosaccharides. His work, carried out in the late nineteenth and early twentieth century was then and is today recognized as one of the outstanding achievements of early structural work.

As there was no way of determining absolute configurations at that time, Fischer derived the configurations of all aldohexoses relative to one enantiomer of glyceraldehyde. He arbitrarily chose the D–configuration for (+)–glyceraldehyde, one of two possibilities, and luckily selected correctly. This was established in 1951 by X–ray diffraction analysis. Consequently, the representations of the molecular structures of monosaccharides used since Fischer, denote not only the correct relative but also the correct absolute configurations.

The formal (HCOH)–group insertions which have led us to the family of aldoses starting with glyceraldehyde, were practically carried out by Fischer applying a method which became known as the Kiliani–Fischer cyanohydrin synthesis. Hydrocyanic acid is reacted with aldehydes and ketones under acid– or base–catalysis to give an isomeric pair of α–hydroxy–nitriles, called cyanohydrins, hydrolysis of which produces α–hydroxy acids.

D-Glucononitrile   D-Arabinose   D-Mannononitrile

Kiliani applied this reaction to aldehydes to obtain an epimeric pair of cyanohydrins and after hydrolysis two epimeric lactones of the aldonic acid, carrying one carbon atom more than the starting aldose. The use of this reaction as a method of ascending the aldose series was made possible by Fischer, who showed that the aldonic acids could be selectively reduced to the corresponding aldoses by sodium amalgam under mildly acidic conditions (pH 3.0–3.5).

In this way it was possible to convert an aldose into two higher diastereomeric aldoses, named epimers, each having one additional hydroxymethylene group between the aldehydic and its adjacent carbon atom. Thus four tetroses, eight pentoses and 16 aldohexoses are obtained from the simplest aldose glyceraldehyde. Since each step gives rise to two products, a chemical method had to be developed to distinguish them. This was done by converting each aldose into a derivative with identical terminal groups and determining whether the products were optically inactive *meso* forms or optically active, thus asymmetric. Therefore, aldoses were treated with strong nitric acid to give the corresponding α,ω–dicarboxylic acids, called aldaric acids. In this reaction, erythrose, ribose, xylose, allose and galactose give the internally compensated *meso* dicarboxylic acids (e.g. tartaric acid from erythrose), whereas from the other aldoses optically active, asymmetric acids are derived. This technique taken together with the structures of the sugars from which they were derived, allows the five named sugars and their epimers to be structurally characterized.

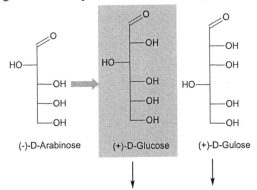

The experimentally known interrelationships which led to the determination of the formerly unknown structure of D–glucose are shaded in gray.

Only glucose, mannose, gulose and idose cannot be assigned structures on this basis. However, Fischer noted that D–glucose and D–gulose give enantiomeric derivatives after appropriate oxidation. There was only one pair of acids having this relationship which could be derived from these four hexoses. It was also known that D–glucose is derived from D–arabinose and consequently glutaric acid must be the oxidation product of D–glucose. Based on this conclusion the structure of D–glucose could be assigned. D–Mannose must be its epimer. The structures of D–gulose and D–idose were similarly deduced. Following this concept, Fischer solved the interrelationships between the monosaccharides like a stereochemical crossword puzzle.

Monosaccharides form cyclic hemiacetals

Monosaccharides exist preferably as cyclic hemiacetals and hemiketals. These arise from the intramolecular nucleophilic attack of a hydroxyl–oxygen atom at the carbonyl carbon atom of the acyclic species. Depending on which hydroxyl group of the monosaccharide chain reacts with the carbonyl group, 5– or 6–membered rings are formed which are called furanoses and pyranoses, respectively, in analogy to their unfunctionalized heterocycle analogs, tetrahydrofuran and tetrahydropyran.

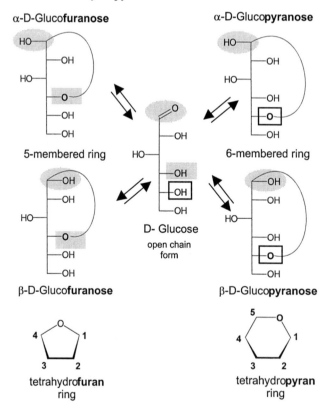

The formation of 5– and 6–membered monosaccharide rings is exemplified by the aldose D–glucose.

From the nucleophilic attack of a hydroxyl group at the prostereogenic carbonyl carbon, two different configurations at C–1 may result in the newly formed sugar ring. Depending on which configuration is created, the hemiacetals are named α– or β–furanoses and α– and β–pyranoses, respectively.

Representation of monosaccharide structures

For the graphic representation of monosaccharide structures, several different semiotic descriptions are used. For the open chain form the Fischer projection is most commonly used, but zig–zag projection is also used. For the description of the hemiacetal forms Fischer projections are not suitable. Cyclic sugars can be represented by the so–called Mills projection, which is the favored way of representing carbohydrate structures in other areas of natural product chemistry, when monosaccharides are part of much larger structures which contain additional non–carbohydrate rings. It is also a favorable representation when synthetic pathways from carbohydrates to chiral non–carbohydrate building blocks have to be described.

In carbohydrate chemistry, the Haworth projection is more common. In the Haworth projection the tetrahydrofuran ring is assumed to be planar and the hydroxyl groups are arranged above and below the plane of the ring. This remains a favorable structural representation of carbohydrates for many purposes. The Haworth projection, however, provides no information about the conformation of carbohydrates. The 6–membered ring of monosaccharides is certainly not planar, but adopts a regular chair conformation and thus, the cyclic forms of carbohydrates are most accurately drawn as chairs. This representation also facilitates the distinction between axial and equatorial ring positions, a difference which is of some importance in carbohydrate chemistry. Conformational representations of sugars are favored in this book.

| Fischer projection | Zig-zag projection | Mills projection | Haworth projection | Chair conformation |

Different projections for the structure of the open chain and the cyclic form of β–D–glucose. They are all identical in terms of their absolute configurations, as can easily be shown for each stereogenic center applying the CIP convention.

To transform a structure given in the Fischer projection into a Haworth or chair representation, the following rule can be followed: A hydroxyl group located left in the Fischer projection 'translates' into an upward position in the Haworth (chair)–representation and a Fischer–right–hydroxyl translates into a Haworth (chair)–down–OH.

The anomeric configuration

The new stereogenic center generated by hemiacetal ring closure is called the anomeric center. The two possible stereoisomers are referred to as anomers, designated as α or β according to the stereochemical relationship between the anomeric center and the configuration of the most distant stereogenic center. If the hydroxyl groups bound to this center point in the same direction (*cis*), this anomer is called the α–anomer, when they are pointing in opposite directions (*trans*), it is named β. Anomers are diastereomers.

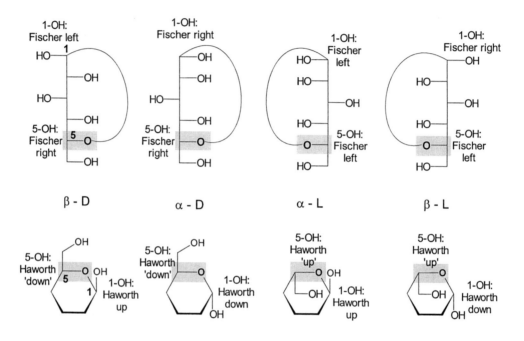

Corresponding structures of α– and β–anomers of monosaccharide hemiacetals of the D– and L–series, represented as Fischer and Haworth projections; the correct numbering of the carbohydrate ring is indicated.

Thus, for D–glucose and all compounds of the D–series, α–anomers have the hydroxyl group at the anomeric center projecting downwards in Haworth formulae; α–L–compounds have this group projecting upwards. The β–anomers have the opposite configurations at the anomeric centers, i.e. the hydroxyl group projects upwards and downwards for β–D– and β–L–compounds, respectively. A wavy line is used for the anomeric bond when the anomeric configuration is not specified.

Assigning the anomeric configuration can sometimes be tricky and might well be exemplified by the pentopyranose arabinose and the similarly configured 6–deoxy sugar fucose.

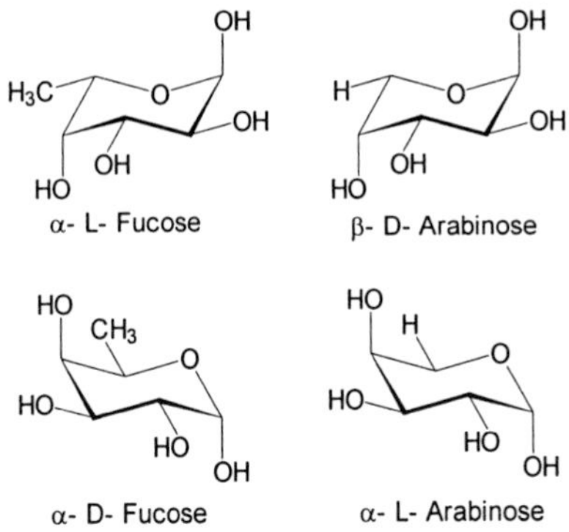

α- L- Fucose   β- D- Arabinose

α- D- Fucose   α- L- Arabinose

Mutarotation

Each crystalline free sugar is a discrete stereoisomer. On dissolution in water however, the hemiacetal ring opens and reforms to give products with different ring sizes and configurations at the anomeric center. This equilibration occurs with all reducing saccharides and is accompanied by a change in optical rotation known as mutarotation. It can be acid– and base–catalyzed.

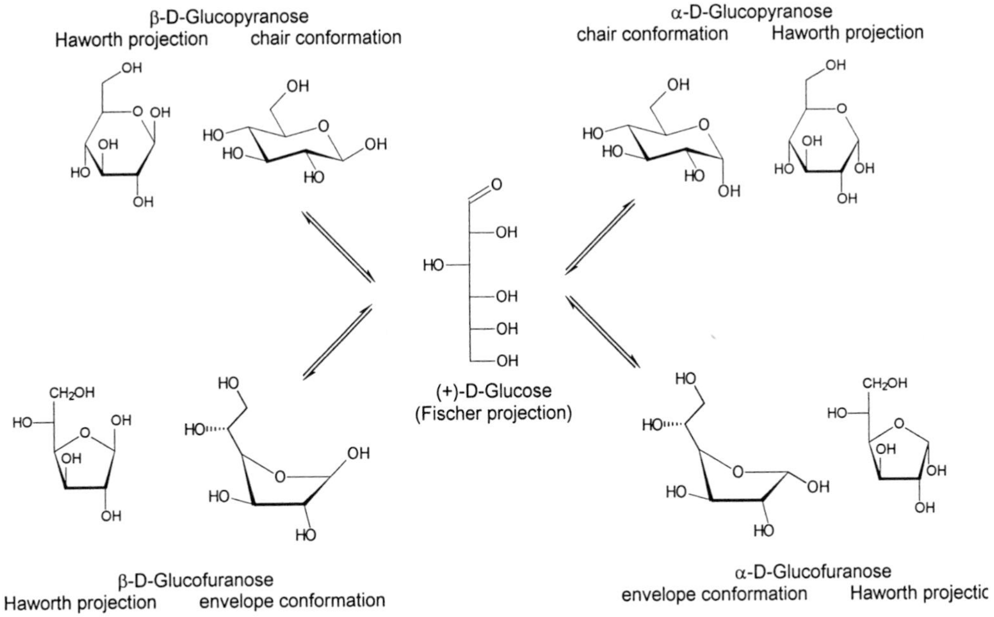

β-D-Glucopyranose
Haworth projection   chair conformation

α-D-Glucopyranose
chair conformation   Haworth projection

(+)-D-Glucose
(Fischer projection)

β-D-Glucofuranose
Haworth projection   envelope conformation

α-D-Glucofuranose
envelope conformation   Haworth projectic

Mutarotation of D–glucose in solution leads to a mixture of α– and β–pyranoses as well as α– and β–furanoses. In addition to the structures shown, several other species can be involved in mutarotation including the acyclic hydrate and even septanoses and oxetanoses.

Thus, the rotation value of a freshly prepared sugar solution changes with time until mutarotation is complete and an equilibrium value has been reached. If either pure α–D–glucopyranose with a specific rotation value of +112° or pure β–D–glucopyranose with a specific rotation value of +19° are dissolved in water, the rotation value keeps changing until an equilibrium value of +52.7° is reached after about 3 hours.

Table 2–1. The percentage compositions of sugars in aqueous solution at equilibrium can be determined using polarimetry and NMR spectroscopy.

Carbohydrate	Temperature (°C)	α–Pyranose (%)	β–Pyranose (%)	α–Furanose (%)	β–Furanose (%)	Carbonyl form	Equilibrium rotation value
Glucose	31	38.0	62.0	0.5	0.5	0.002	+53° (D)
Mannose	44	65.5	34.5	0.6	0.3	0.005	+15° (D)
Galactose	31	30.0	64.0	2.5	3.5	0.02	+80° (D)
Rhamnose	44	65.5	34.5	0.6	0.3	0.005	+9° (L)
Fructose	31	2.5	65.0	6.5	25.0	0.8	-92° (D)
Xylose	31	36.5	63.0	0.3	0.3	0.002	+19° (D)
Ribose	31	21.5	58.5	6.4	13.5	0.05	-24° (D)

Conformations of monosaccharides

Like cyclohexane, the 6–membered ring of monosaccharides also exists in two isomeric chair conformations, which are specified as $^1C_4$ and $^4C_1$, respectively, where the letter C stands for 'chair' and the numbers indicate the carbon atoms located above or below the reference plane of the chair, made up by C–2, C–3, C–5 and the ring oxygen.

The conformational shape of a pyranose is mainly governed by the relative stability of the two possible chair conformations which are both free of torsional strain, but one of which, in most cases, is clearly energetically unfavored because of van der Waals interactions of the ring substituents. Thus, the $^1C_4$ conformation of β–D–glucopyranose is clearly unfavored compared to its $^4C_1$ conformation because the van der Waals repulsion of the 1,3–diaxially positioned ring substituents result in a free energy difference between the two chairs of approximately 25 kJ/mol. Consequently only one, the $^4C_1$ conformation of β–D–glucose is observed by NMR spectroscopy. On the other hand, as expected from their configurations, the energy difference between both chair conformations of α–idose and α–altrose is so small, that consequently both forms can be observed in the NMR spectrum.

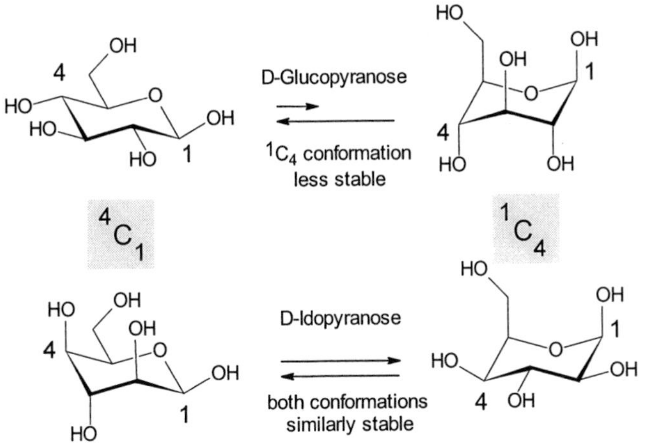

When chair representations are turned around for graphical reasons, one has to be careful not to accidentally change the absolute stereochemistry. This can sometimes be tricky. For example, the mirror image obtained from D–galactose is correctly represented in a $_1C^4$ conformation, however, more conventionally it is drawn in the form of the stereochemically synonymous $^1C_4$ chair.

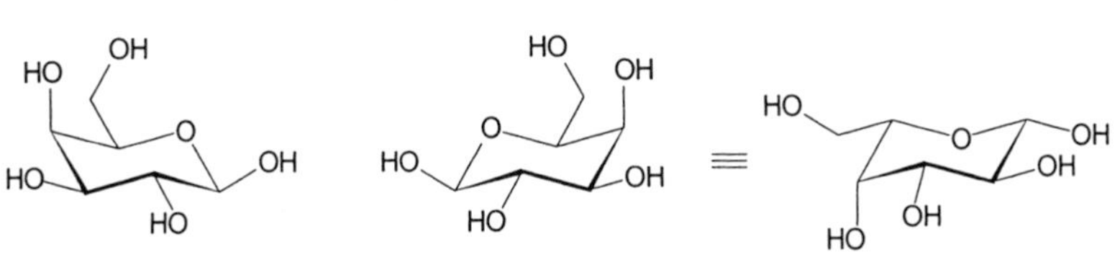

β-D-Galactose   β-L-Galactose

Minor conformational isomers can be of importance, when they form the starting material in chemical reactions which do not proceed with the major conformer. An important example is the synthesis of anhydro derivatives of monosaccharides which are versatile intermediates in carbohydrate chemistry. Treatment of 6–*O*–tosyl–glucopyranose with sodium methoxide gives the 1,6–anhydro derivative which results from nucleophilic attack of the anomeric hydroxyl group at the 6–position of the sugar ring in the unfavored $^1C_4$ conformation! Treatment of the analogous methyl glucoside, where the anomeric hydroxyl group is no longer available, with sodium hydride and benzyl bromide in DMF gives the benzylated 3,6–anhydro glycoside, also in the $^1C_4$ conformation.

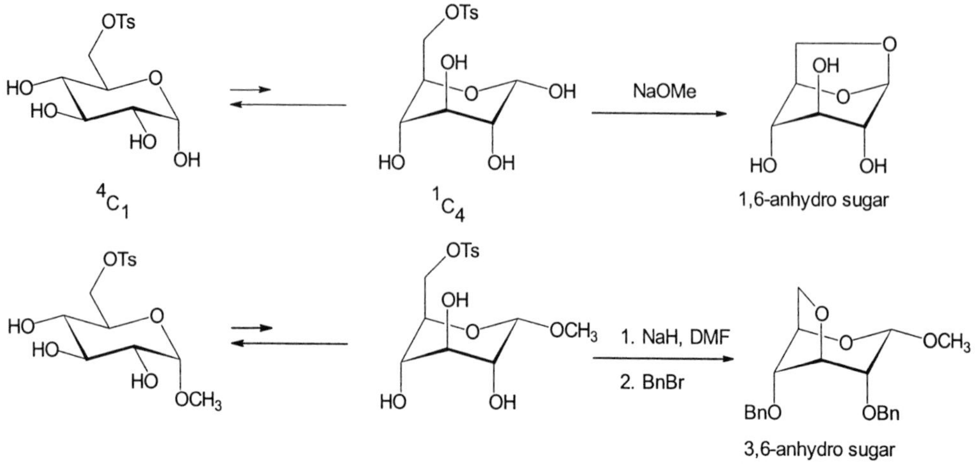

OTs
HO
HO
O
HO
OH
$^4C_1$
OTs
OH
O
OH
HO
OH
$^1C_4$
NaOMe
O
OH
O
HO
OH
1,6-anhydro sugar
OTs
HO
HO
O
HO
OCH3
OTs
OH
O
OCH3
HO
OH
1. NaH, DMF
2. BnBr
O
O
OCH3
BnO
OBn
3,6-anhydro sugar

Other principal conformations of pyranoses are half–chair (H), boat (B), and skew (S) conformation, which are named as indicated. The chair is by far the most stable and only the skew conformation has an energy minimum in a similar range, but this is still some 20 kJ higher than the chair. Principal conformations of the furanose ring are the envelope forms (^{1}E, E_1, ^{2}E, E_2, ^{3}E, E_3, ^{4}E, E_4, ^{O}E, E_O) and the twist forms ($^{O}T_1$, $^{1}T_O$, $^{1}T_2$, $^{2}T_1$, $^{2}T_3$, $^{3}T_2$, $^{3}T_4$, $^{4}T_3$, $^{4}T_O$, $^{O}T_4$).

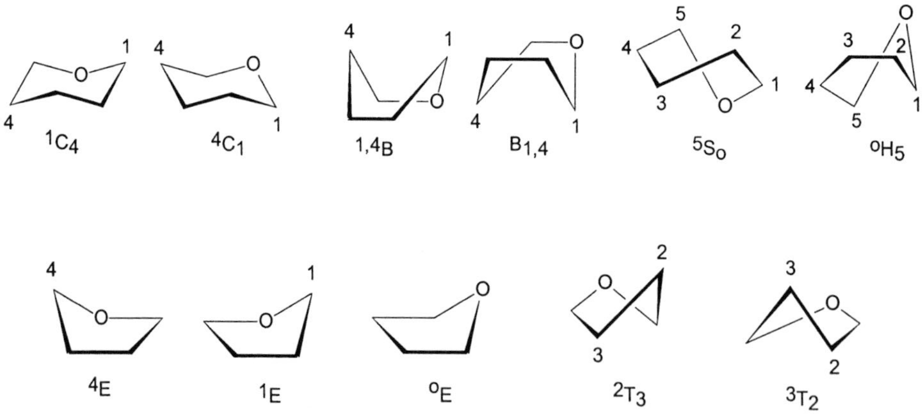

$^1C_4$
$^4C_1$
$^{1,4}B$
$B_{1,4}$
^5S_O
$^OH_5$
^4E
^1E
^OE
$^2T_3$
$^3T_2$

In addition to intramolecular van der Waals interactions, carbohydrate conformations are determined by some other factors, such as electrostatic interactions as well as intramolecular hydrogen bond formation and especially the anomeric effect.

The anomeric effect

The equatorially positioned substituents of a carbohydrate ring are, for steric reasons, the most energetically favored, compared to their axial counterparts, as is the case in every molecule with a chair conformation. However, the anomerically bound groups in carbohydrates do not follow this rule completely: For D–pyranoses, D–pyranosides and especially carbohydrate derivatives with electronegative groups at the anomeric center the anomeric α–configured derivatives with the anomeric group located in an axial position are often more stable than would be predicted from the steric interactions they have with adjacent substituents. An aqueous solution of D–glucose, for example, contains the α– and the β–form in a ratio of 36:64, and the effect is even pronounced for D–mannose, where the α:β–ratio is 69:31. The unusual preference of the sterically unfavored axial position over the equatorial position at the anomeric center has been termed the 'anomeric effect' by R. Lemieux.

The anomeric effect was discovered in the case of carbohydrates but has been found to be of general importance for molecules, where two heteroatoms are bound to a tetrahedral center. Thus, the essential group for the appearance of an anomeric effect is

$$-C-Y-C-X \quad \text{where} \quad Y = N, O, S$$
$$\text{and} \quad X = Br, Cl, F, N, O, S$$

The anomeric effect can be explained in several ways. It partly involves a dipole-dipole effect based on intramolecular electrostatic interactions of two dipoles next to the anomeric center. One of the two dipoles arises from the two lone electron pairs of the endocyclic carbohydrate ring oxygen. The other dipole points along the polarized bond between the anomeric carbon atom and its bound atom X. Anomeric configurations, where the two dipoles partially neutralize each other are favored over the diastereomers where the anomeric configuration leads to partial intramolecular addition of the two dipoles. The anomeric effect virtually ensures the axial configuration of an electronegative substituent at the anomeric center such as in the case of acetobromoglucose, where the β–anomer is unknown.

Unstable anomer   Anomer which is favored by the anomeric effect   partial dipole moments

Also unfavorable lone pair–lone pair interactions have been used to explain the anomeric effect. Most importantly however, the anomeric effect is a stereoelectronic effect, in which a lone pair of electrons located in a n–molecular orbital of the atom Y overlaps with the antibonding σ*–orbital of the C–X bond. This favorable $n_{Y\rightarrow}\sigma^*$ delocalization of nonbonding electrons ('negative hyperconjugation') is only possible with an anti-periplanar arrangement of the involved orbitals as found in the axial anomer. This interaction is also reflected by bond length changes, slightly shortening the Y–C1 bond while lengthening the C1–X bond.

n orbital

σ* orbital

Favored anomer due to $n_{Y\rightarrow}\sigma^*$ delocalization of nonbonding electrons which is possible with an anti–periplanar arrangement of the involved orbitals. A significant effect is obtained when the σ_{CX}^* orbital is of low energy.

The anomeric effect is of a different size for every specific case. It is strongly influenced by the substituent at C–2. When this is equatorial, as in glucose and galactose, the anomeric effect is weakened, and is enhanced in the case of an axial C–2–substituent as in mannose. Moreover, the nature of the anomeric group is of crucial influence for the anomeric effect, as it is proportional to the electronegativity of the anomerically bound atom. Solvents also influence the anomeric effect, such that increased polarity of the solvent used decreases the influence of the anomeric effect on the equilibration of the two alternative conformers in solution.

The anomeric effect may even lead to conformational changes as for β–xylopyranosyl bromide which prefers the sterically unfavored $^1C_4$ conformation due to the strong anomeric effect of the bromo atom.

If the substituent at the anomeric center is clearly electropositive compared to the anomeric carbon, such as a positively charged nitrogen atom, the same electrostatic considerations as stated for the anomeric effect lead to the stabilization of the anomer with the equatorially positioned anomeric group. This effect, which is causally no different from the anomeric effect, has been termed the 'reverse anomeric effect'. It is assisted by the fact that an equatorial ring position is energetically favored due to steric reasons, especially in the case of a large substituent like a pyridinium group.

An anomeric pyridinium group, for example, leads to a reverse dipole at the anomeric carbon and consequently a 'reverse anomeric effect' is observed.

In alkyl glycopyranosides the anomeric effect operates not only along the endocyclic C–1 oxygen bond but also along the exocyclic C–1 oxygen bond. The anomeric effect leading to prefered conformations of the exocyclic alkoxy group is called 'exo-anomeric effect'. Again it is an anti-periplanar arrangement of a lone pair on the aglycon oxygen and the C1–O5 bond which determines the favored conformation. In axially configured acetals the exo-anomeric effect is less important because it operates in the opposite direction than the 'endo-anomeric effect'. However, in an equatorial acetal the exo-anomeric effect is dominant and dictates the prefered conformation of the aglycon alkoxy group.

Modified monosaccharides

Standard monosaccharides are widely distributed in nature and easily obtained by degredation of polysaccharides or isolation from various materials for example plants, microorganisms, yeast, marine sources or milk.

However, the structures of many sugars which are known today are different from the general formula $C_n(H_2O)_n$. They are modified by amino groups or deoxygenation to name but a few. Many of them have been given trivial names as they are frequently found as constituents of different classes of biologically active compounds such as hormones, alkaloids, antibiotics, or cardiac glycosides.

Glucosamine
(2-Amino-2-deoxy-D-glucopyranose)

Galactosamine
(2-Amino-2-deoxy-D-galactopyranose)

Daunosamine
(3-Amino-2,3,6-trideoxy-
L-*lyxo*-hexopyranose)

L-Fucose
(6-Deoxy-L-galactopyranose)

L-Rhamnose
(6-Deoxy-L-mannopyranose)

Quinovose
(6-Deoxy-D-glucopyranose)

Olivose
(2,6-Dideoxy-
D-*arabino*-hexopyranose)

Digitoxose
(2,6-Dideoxy-
D-*ribo*-hexopyranose)

Cymarose
(2,6-Dideoxy-3- O-methyl-
D-*ribo*-hexopyranose)

Alditols

Mild reduction of aldoses and ketoses with, for example, sodium boronate leads to sugar alcohols which are called alditols. Like the corresponding monosaccharides they are named according to the number of carbon atoms, as pentitols and hexitols, for example. Whereas in the reduction of aldoses only one product is obtained, reduction of ketoses leads to two diastereomers.

Reduction of D–altrose gives the same product (D–altritol) as D–talose in the same reaction; also D–glucose and D–gulose give the same alditol, D–glucitol, which is sometimes also called sorbitol. Consequently, only six different D–alditols are obtained from the eight stereoisomeric D–aldoses. From those D–allitol and D–galactitol are *meso* forms, the other four are chiral. Alditols are, like the aldaric acids, sugar derivatives with identical end groups and therefore include interesting symmetry elements. This may be utilized to convert, through a series of chemical conversions, a sugar from the D–series into an L–sugar.

The six diastereomeric D–alditols derived from aldohexoses or aldoketoses, respectively.

Cyclitols

Cyclitols as cyclic polyhydroxyalkanes have a lot in common with monosaccharides and are often used in carbohydrate chemistry. Of the cyclitols the hexahydroxycyclohexanes are called inositols and are best known. All nine possible stereoisomers have been prepared. *myo*–Inositol is often simply referred to as inositol. It occurs widely in nature, particularly as a phosphate ester.

Closely related to the inositols are the carba–sugars which were previously known as pseudo–sugars. They differ from normal sugars in the replacement of the ring oxygen atom with a methylene group.

Structures of all nine stereoisomeric inositols; only the *chiro*–inositols are chiral, the other stereoisomers are *meso* forms. All inositols are numbered as indicated for *cis*–inositol.

Nomenclature of monosaccharides

Nomenclature of carbohydrate derivatives is dependent on the trivial names which have been given to the most common aldoses consisting of three, four, five and six carbon atoms. The basis for naming a carbohydrate is the parent monosaccharide in its acyclic form from which the various configurational prefixes are derived; these are included in italics in the name of the sugar. When a monosaccharide contains a deoxy function, this chain position is ignored for the purpose of assigning a configurational prefix. Thus, the configurational prefix covers the entire sequence of chiral centers even when a CH_2 group divides the chiral centers into two sets. The prefix 'deoxy' is used preceded by the position. For deoxy sugars, however, trivial names are also commonly used, especially when the deoxygenation is not at a chiral center.

4-O-Methyl-D-xylitol   3-Deoxy-D-*ribo*-hexose

Modification of a hydroxyl group or its replacement by another function is envisaged as substitution. The stereochemistry at the carbon atom carrying the new substituent is expressed as before with the substituent regarded as equivalent to OH. The substituted position is indicated as 'deoxy–(substituent–prefix)' in the name of the derivative preceded by the position of the modification, e.g. 4–deoxy–4–methyl or 3–bromo–3–deoxy.

For assigning the configurational symbols D and L, the configuration at the highest–numbered center of chirality is used. This stereogenic carbon atom is called the 'configurational atom'. To specify the anomeric configuration an 'anomeric reference atom' is used. This is the same as the configurational reference atom, unless multiple configurational prefixes are used.

This is the case when monosaccharide chains consist of more than six carbon atoms (e.g. aldoheptoses and aldooctoses). These have not been given trivial names, but are systematically named by adding two (or more) configurational prefixes to the stem name, as derived from the names of the lower aldose members. Prefixes are assigned in order to the chiral centers in groups of four, beginning with the group proximal to C–1. The prefix relating to the group of carbon atom(s) farthest from C–1 (which may contain less than four atoms) is cited first. Each group of asymmetrically substituted atoms represented by a particular prefix has its own configurational symbol, specifying the configuration of the highest numbered atom of this group as D or L.

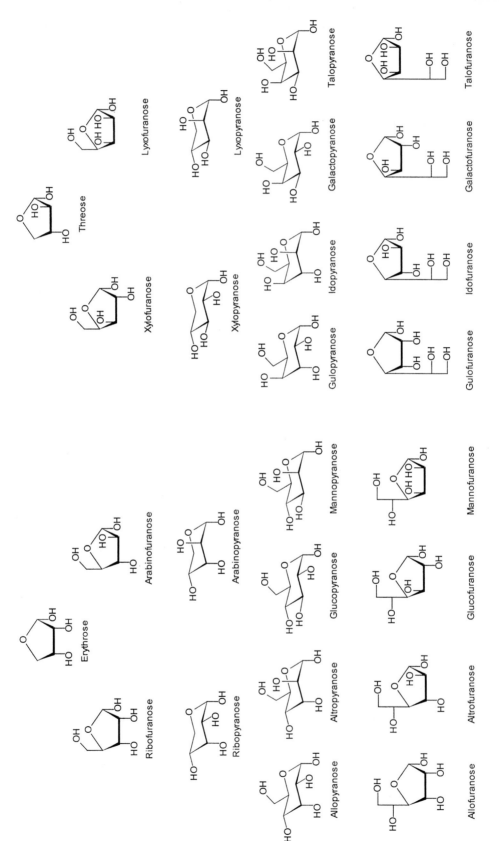

Cyclic forms of the α–D–aldoses.

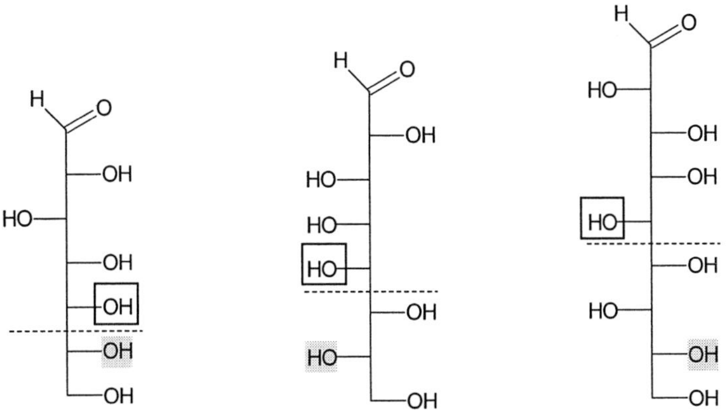

D- *glycero* -D- *gluco*-Heptose   L- *threo* -L- *altro*-Octose   D- *xylo* -L- *galacto*-Nonose

In cases where more than one configurational prefix has to be used for naming a mono-saccharide, the configurational reference atom can no longer be regarded as the anomeric reference atom. In this case, the position of the hydroxy group closest to the stereogenic center which is involved in the ring formation, forms the basis for choosing the anomeric prefix.

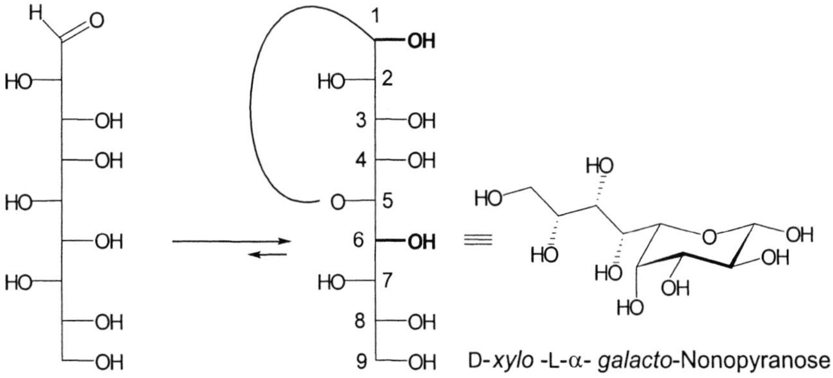

D- *xylo* -L- *galacto*-Nonose

D- *xylo* -L- α- *galacto*-Nonopyranose

Ring formation of D–xylo–L–galacto–nonose to form the α–pyranose. Note that C–6 is the anomeric reference atom.

There are quite a number of examples where more than an aldehydic or carbonyl group is embedded in a monosaccharide, as for example in ketoaldonic acids. Names for ketoaldonic acids are formed by replacing the ending 'ulose' of the corresponding ketose by 'ulosonic acid', preceded by the location of the ketonic carbonyl group. Numbering starts at the car-boxy group. An important example is the IUPAC name for the sialic acid *N*–acetyl–

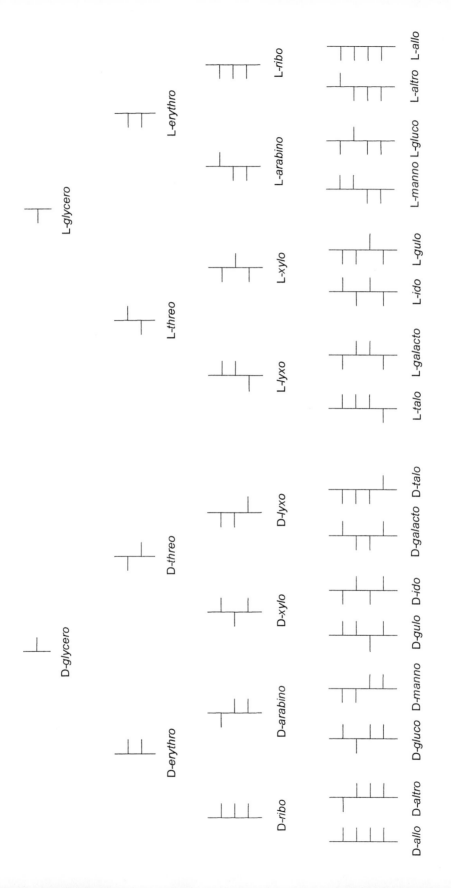

The configurations and respective prefixes of the D– and L–series, on which carbohydrate nomenclature is dependent. The lowest numbered atom is positioned up the chain in the Fischer projections used, numbering proceeds down the chain. Epimeric pairs are positioned side by side.

neuraminic acid, which is abbreviated to Neu5Ac, 5–acetamido–3,5–dideoxy–D–glycero–
α–D–galacto–non–2–ulopyranosonic acid.

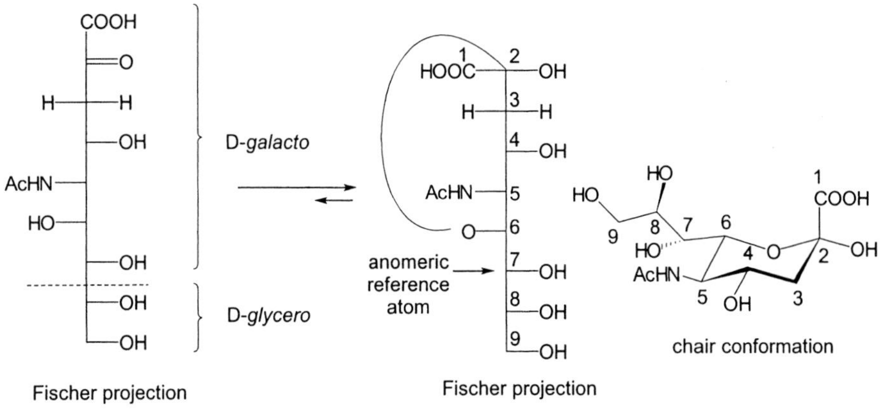

Fischer projection   Fischer projection

The structure of Neu5Ac is given as a Fischer projection; ring formation in the α–mode is
indicated in the Fischer projection and in the chair conformation.

   Naming of carbohydrates can become really sophisticated. In avoiding the complicated
IUPAC nomenclature in such cases, trivial names are frequently used. One of the typical
examples is the naming of unsaturated monosaccharides. The group of 1,2–unsaturated
monosaccharides, commonly referred to as 'glycals' are named 1,5–anhydro–2–deoxy–1–
enitols according to IUPAC and a 2,3–unsaturated derivative is called a 2,3–dideoxy–2–
eno–pyranose.

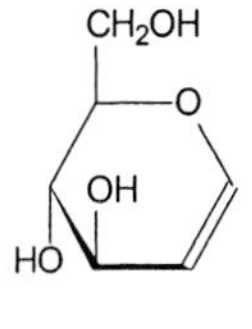

trivial name: D-glucal
1,5-Anhydro-2-deoxy-D- *arabino* -hex-1-enitol   2,3-Dideoxy-α-D- *erythro* -hex-2-enopyranose

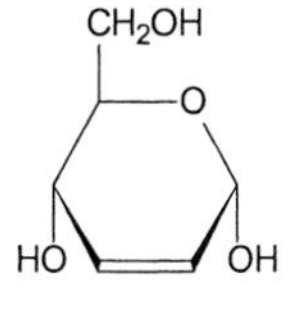

   Furthermore, its worthwhile to note, that monosaccharides in which an alcoholic hydroxyl group has been replaced by an amino group are called *amino sugars*, monosaccharides carrying the amino group at the anomeric center are named *glycosyl amines*. Mono-
carboxylic acids formally derived from aldoses by replacement of the aldehydic group by a
carboxy group are called *aldonic acids*, those derived from aldoses by replacement of the
terminal CH_2OH group with a carboxy group are named *uronic acids*. The dicarboxylic acids formed from aldoses by replacement of both terminal groups (CHO and CH_2OH) with
carboxy groups are called *aldaric acids*.

2.2   Structure of simple oligosaccharides

When a free, so–called reducing monosaccharide, which resembles a cyclic hemiacetal in its cyclic form, reacts with another monosaccharide under acid catalysis a mixed acetal is formed, which is generally called a glycoside and named a disaccharide in the simplest case. The bond between the two components of the acetal is called the glycosidic bond. Thus, monosaccharides can react as monomers to form oligo– and polymers, called oligosaccharides and polysaccharides, respectively. Indeed, most carbohydrates do occur as oligo– and polysaccharides, rather than as monosaccharides.

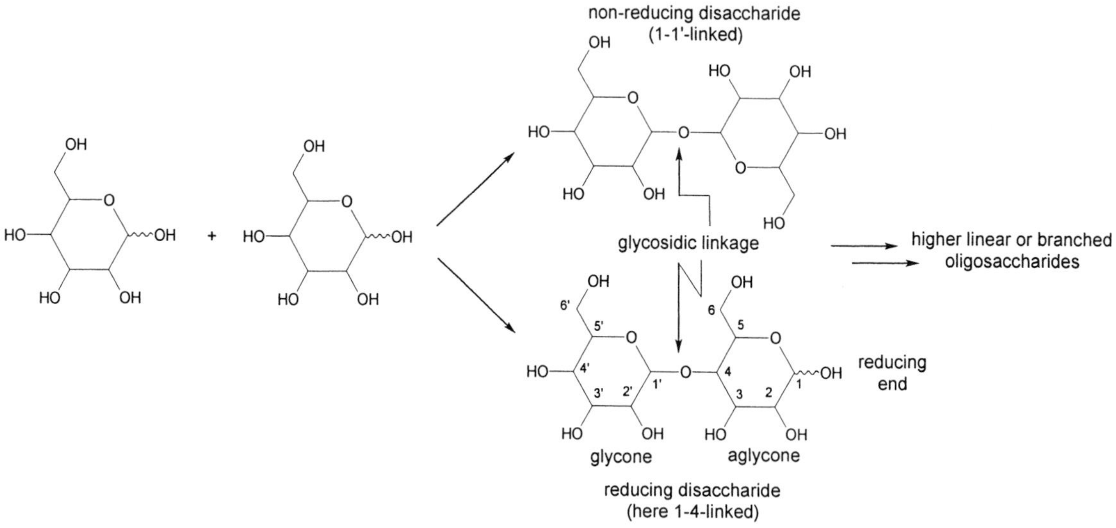

Linking monosaccharides via glycosidic bonds leads to the formation of oligosaccharides, which may be linear or branched, reducing or non–reducing. The moiety at the reducing end is called the aglycone, the remaining oligosaccharide portion the glycone part. Numbering of oligosaccharide carbon atoms use numbers 1 to 6 for the reducing end residues and primed and double–primed numbers, respectively, for the glycone constituents. Though even triple–priming is used for higher oligosaccharides an individual solution is often sought for labeling of the atoms of complex products.

According to IUPAC nomenclature, oligosaccharides are compounds in which monosaccharide units are joined by glycosidic linkages. Depending on the number of units, they are called disaccharides, trisaccharides, tetrasaccharides, etc. There is no strict borderline drawn with polysaccharides, however the term 'oligosaccharide' is commonly used to refer to a defined structure as opposed to a polymer. Oligosaccharides in which structural elements such as the interglycosidic linkages are of other types than those occurring in the natural examples are often called oligosaccharide analogs.

Systematic nomenclature of oligosaccharides produces rather lengthy names which are often abbreviated in the form of a short–cut nomenclature. For the common di– and trisaccharides trivial names are normally used instead of the official ones. The non–reducing elements of an oligosaccharide is given the suffix '–yl'. In the case of non–reducing oligosaccharides the monomer constituents are listed in alphabetical order. Thus lactose is called β–D–galactopyranosyl–(1→4)–D–glucopyranose, abbreviated Gal*p*–β(1,4)–Glc (the italic '*p*' indicates the pyranose form); and sucrose is named α–D–glucopyranosyl–β–D–fructofuranoside (Glc*p*–α(1,2)–β–Fru).

Disaccharides are the simplest oligosaccharides and a number of them belong to the least expensive carbohydrates available as they can be easily obtained by hydrolysis of abundant polysaccharides (often regarded as 'renewable' resources). Thus maltose is obtained in large quantities from starch, cellobiose from cellulose. α,α–Trehalose, in which two glucose moieties are linked via both of their anomeric centers, is a non–reducing sugar and is found in yeasts and fungi. Lactose is naturally occurring in milk of mammals; it serves as the principal source of energy for their young. Sucrose (in German *'Saccharose'*) is extracted from sugar cane. Sucrose is chiefly found in sugar cane and sugar beet, but is also found in many other plants, especially fruits. Its particular attribute is its sweetness. Although many carbohydrates are sweet, sucrose has become the sugar of commerce, primarily because of the ease by which it is obtained in high purity from sugar cane or sugar beets.

Also trisaccharides occur relatively often in nature whereas larger oligosaccharides are rarely found in the free form. On the other hand, complex oligosaccharides in numerous structural variations are an important part in many biologically active natural products and moreover are linked to proteins or lipids forming glycoconjugates which are of major importance in cell biology (cf. chapters 6 and 7).

Many oligosaccharides and oligosaccharide analogs are known for their biological effects. For example, a large number of carbohydrates has been evaluated as inhibitors of the enzymes which are involved in polysaccharide digestion, the so–called glycosidases. A large number of α–glucosidase inhibitors were originally isolated from various species of *Streptomyces* bacteria. Among those, acarbose has become an important antidiabetic drug, as it serves as an excellent α–amylase inhibitor. It is a carba–oligosaccharide, also called pseudo–oligosaccharide, consisting of two glucose moieties, a 4–amino–4,6–dideoxy–glucose derivative and a tetra–hydroxylated cyclohexene derivative.

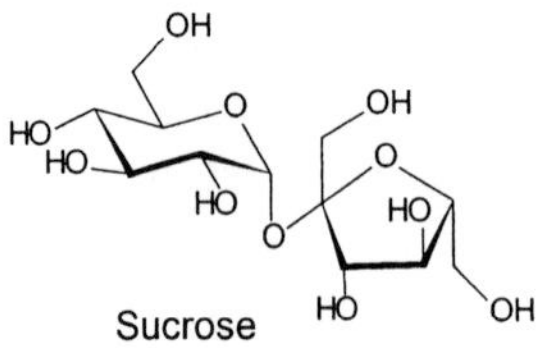

Sucrose
Glc(α1-β2)Fru

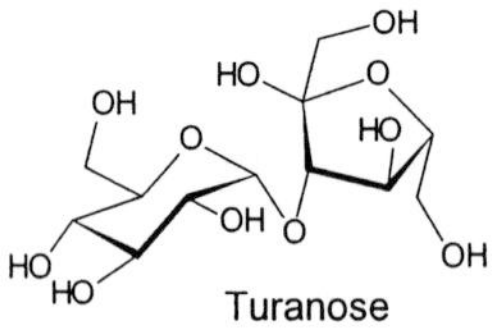

Turanose
Glc(α1-3)Fru

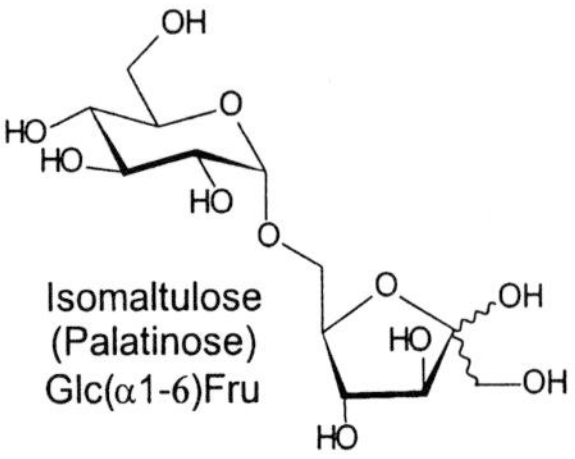

Isomaltulose
(Palatinose)
Glc(α1-6)Fru

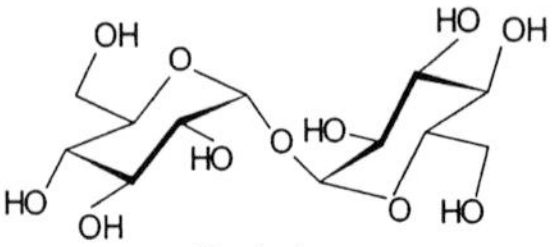

Trehalose
Glc(α1-α1)Glc

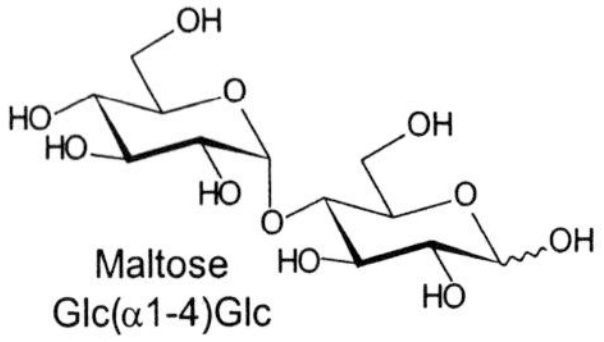

Maltose
Glc(α1-4)Glc

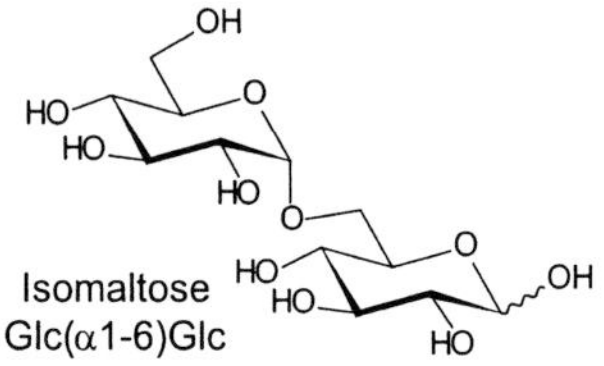

Isomaltose
Glc(α1-6)Glc

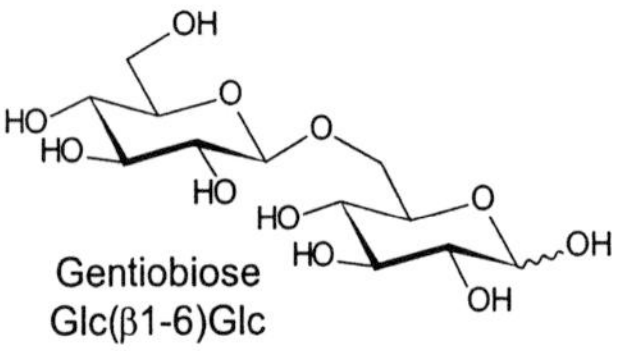

Gentiobiose
Glc(β1-6)Glc

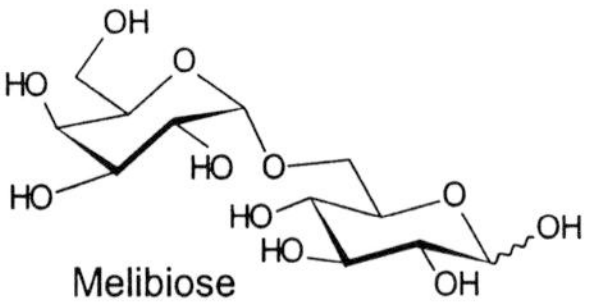

Melibiose
Gal(α1-6)Glc

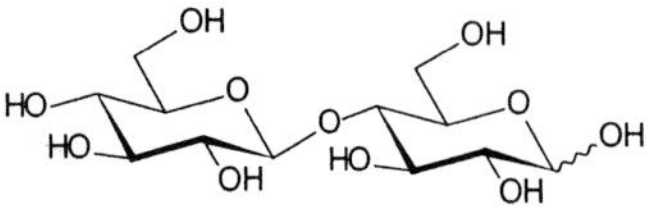

Cellobiose
Glc(β1-4)Glc

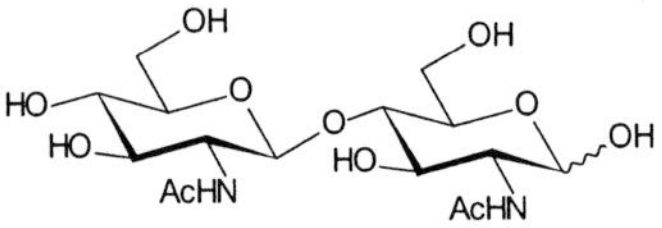

Chitobiose
GlcNH₂(β1-4)GlcNH₂

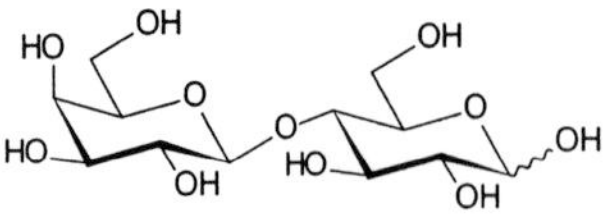

Lactose
Gal(β1-4)Glc

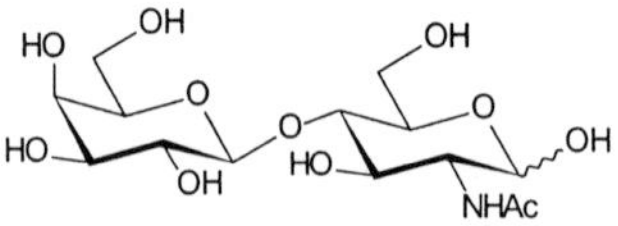

N-Acetyllactosamine
Gal(β1-4)GlcNAc

Structures and trivial names of the most common disaccharides.

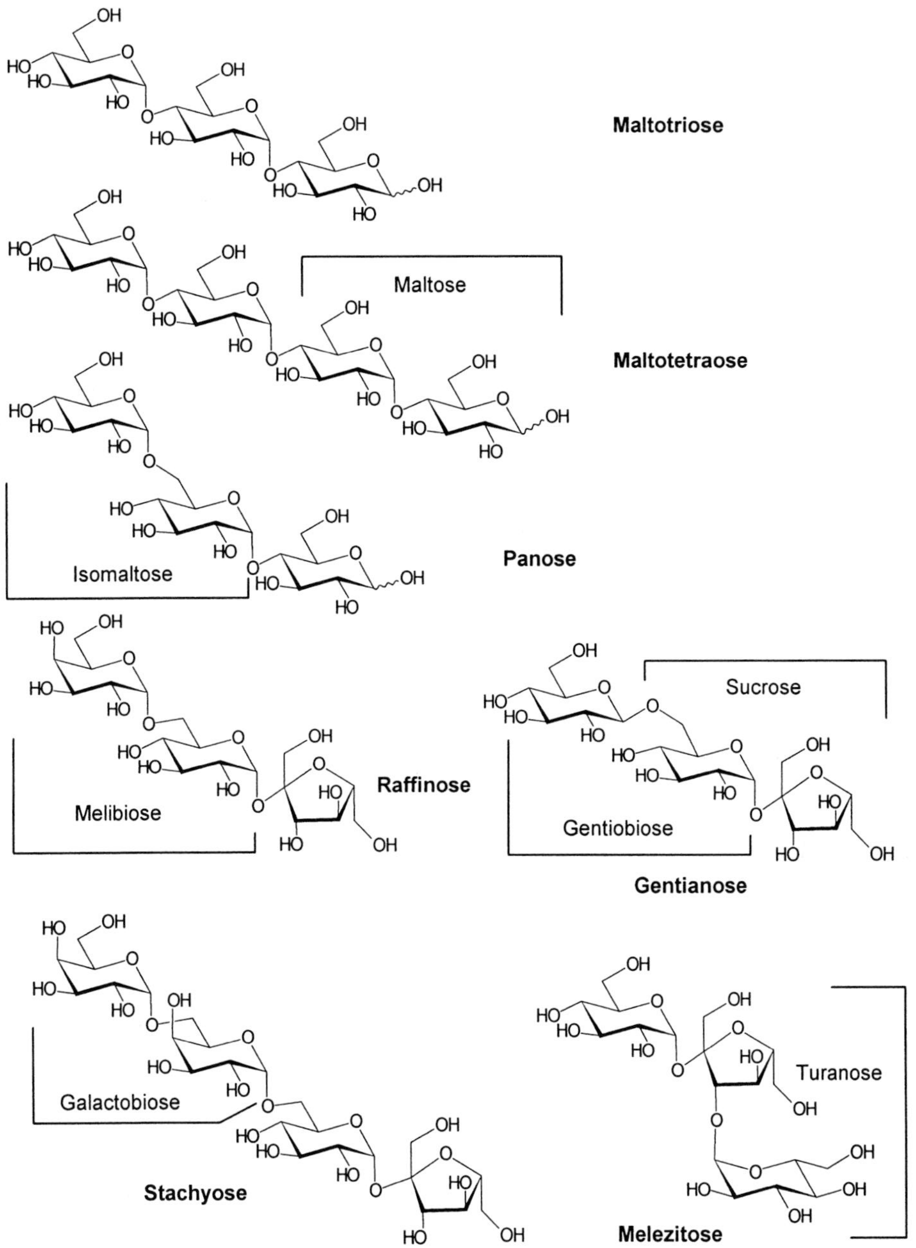

Maltotriose

Maltotetraose

Maltose

Panose

Isomaltose

Melibiose

Raffinose

Sucrose

Gentiobiose

Gentianose

Galactobiose

Stachyose

Turanose

Melezitose

Structures and trivial names of common oligosaccharides found in nature. Names of disaccharide sub–structures are also indicated.

The first oligosaccharide analog with pharmaceutical relevance, streptomycin, was dis-covered in 1944 as a chemotherapeutic agent and has been used to treat some penicillin–resistant strains of Gram–negative and Gram–positive bacteria including mycobacteria. Streptomycin and its relatives were found in culture filtrates of several *Streptomyces* strains. The middle furanose part is called streptose and was the first branched–chain sugar obtained from microorganisms.

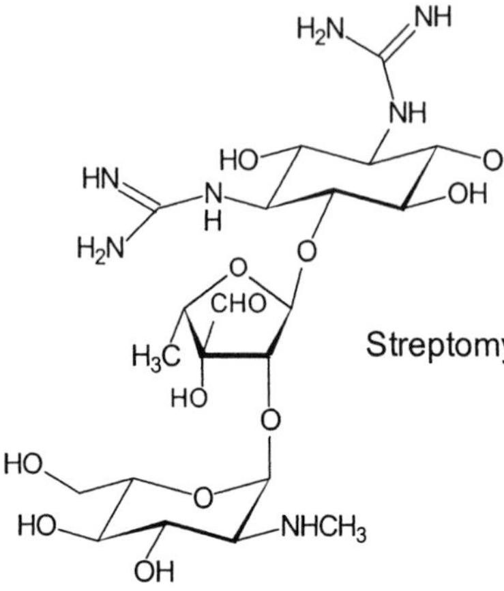

Streptomycin

Many complex oligosaccharides are contained in human milk. They often carry L–fucosyl residues and are believed to have an immunostimulating effect. 2–*O*–Fucosyllactose is a typical example, however, many more complex ones have been isolated from human milk and classified into structural subgroups of which basically seven have been assigned so far.

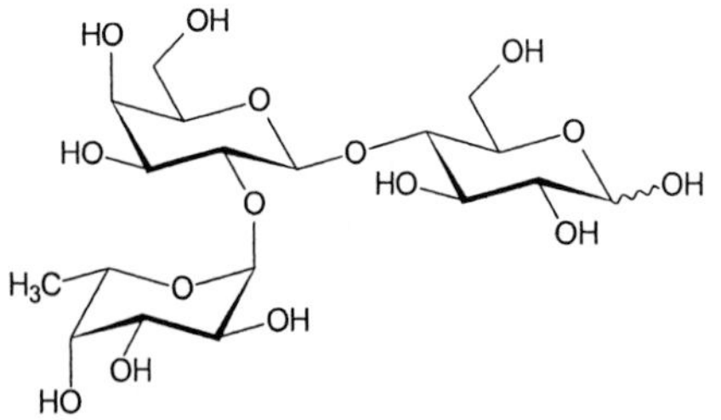

4-*O*-[L-Fuc-α-(1→2)-β-D-Gal]-(1→4)-D-Glc
(2-*O*-Fucosyllactose)

Conformational properties of oligosaccharides

The three–dimensional properties of oligosaccharides are of particular importance in the biochemical context of molecular recognition processes in which carbohydrates are involved. When investigating the shapes of oligosaccharides their energy minimum conformations, both in free and in receptor–bound form have to be addressed, as well as the molecular dynamics of the respective molecules, i.e. are they flexible or rather rigid structures and in which conformational variability do they occur?

In the case of crystalline carbohydrates their three–dimensional structure can be determined by X–ray analysis; for the study of saccharide conformations in solution, NMR experiments on the one hand and theoretical, computer–assisted methods, called molecular modeling, on the other are valuable tools.

For defining oligosaccharide conformations, which are mainly governed by the spatial disposition of the involved glycosidic linkages, a nomenclature was defined using dihedral angles at or close to the respective glycosidic linkage. The two dihedral angles Φ (phi) and Ψ (psi) are used to assign the stereochemistry of a glycosidic link in cases where two secondary alcohols are encountered, i.e. 1→1, 1→2, 1→3, or 1→4 linkages in hexopyranoses. The Φ angle is defined by the H1'–C1'–O–C(aglycone) fragment, Ψ is given by the arrangement of the C1'–O–C(aglycone)–H(aglycone) fragment. In the case of a 1→6 bond the additional degree of freedom is described by the ω (omega) angle, which is defined by the atoms O6–C6–C5–O5. The three most common comformations of the C5–C6 fragment are the staggered conformations which are termed *gg* (*gauche–gauche*), *gt* (*gauche–trans*), and *tg* (*trans–gauche*).

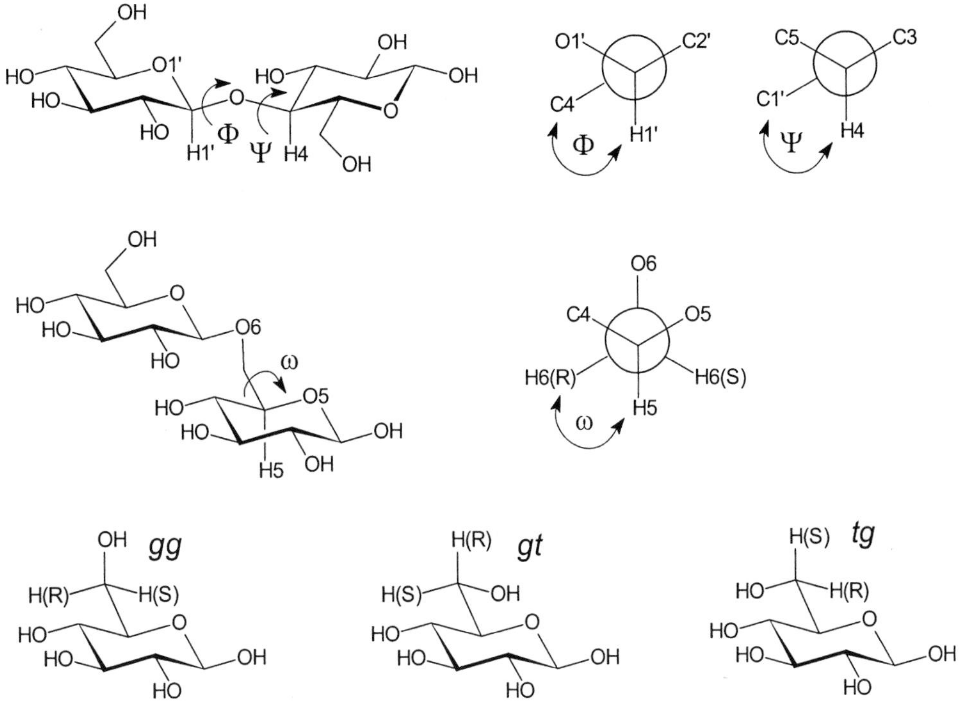

Cyclodextrins

Cyclodextrins are cyclic oligosaccharides, where monosaccharide units form a ring resulting in intersaccharide α–1,4–glycosidic bonds. They are readily available from starch by large scale preparation using cyclodextrin glucotransferases (CGTases), amylolytic enzymes produced by *Bacillus macerans* and other bacterial microorganisms. By this enzymatic procedure a mixture of cyclodextrins is obtained which can be separated by chromatography or fractionated crystallization. The most common and commercially available cyclodextrins are those consisting of six, seven and eight glucose moieties, which are called α–, β, and γ– cyclodextrin (also cyclohexaamylose, cycloheptaamylose and cyclooctaamylose) according to a special nomenclature that this rather unique class of molecules has acquired.

The shapes of cyclodextrins are well established on the basis of X–ray crystallographic studies which reveal that in the solid state, cyclodextrins are conical molecules with well– defined cavities, having hydrophilic exteriors and more hydrophobic interiors. The chemistry which can be performed with cyclodextrins is largely influenced by the remarkable difference in reactivity of the three different types of hydroxyl groups present in the molecule, which decreases in the order: 6–OH > 2–OH > 3–OH.

With the ability to bind other compounds as guest molecules in close proximity to specific hydroxyl groups, cyclodextrins have been perceived as simple models of enzymes and have been found to exhibit catalytic activity in certain reactions, for example in the base– catalyzed hydrolysis of carboxylic acid esters. Cyclodextrins may also be used to bring hydrophobic compounds into aqueous solution or may eventually be developed as drug delivery systems.

Structures of α–, β–, and γ–cyclodextrin together with a common used abbreviation cartoon which also symbolizes the conical shape of cyclodextrins.

2.3   Structure of polysaccharides

It is estimated that approximately 4×10^{11} tons of carbohydrates are biosynthesized each year on earth by plants and photosynthesizing bacteria. The majority of these carbohydrates are produced as polysaccharides. Polysaccharides are macromolecules consisting of a large number of monosaccharide residues. They are sometimes also called glycans. (The term 'glycan' is also used for the saccharide component of a glycoprotein even though the chain length may not be large; cf. chapter 6). For polysaccharides which contain a substantial proportion of amino sugar residues the term glycosaminoglycan is a common one. Polysaccharides which consist of only one kind of monosaccharide are called homopolysaccharides (homoglycans); when they are built up of two or more different monomeric units they are named heteropolysaccharides (heteroglycans). In the latter type, the monosaccharide units are usually linked to each other in a definite pattern, rather than randomly. Certain sequences of monomeric building blocks are often found to be regularly repeated as so–called repeating units. Homo- as well as heteropolysaccharides can be linear or branched.

For polysaccharides, as for every polymer, it is not possible to attribute one distinct molecular weight as they are polydisperse molecules, which are characterized by an average molecular weight. The number of monosaccharide units in a polysaccharide is termed degree of polymerization or d.p. The size of a polysaccharide varies between approximately 16,000 and 16,000,000 daltons (Da).

Polysaccharides exist in an enormous structural diversity as they are produced by a geat variety of species, including microbes, algae, plants and animals. Among these are fructans, xanthans, fucans, bacterial gel polysaccharides, capsule polysaccharides of bacteria, or agar, which is a mixture of two polysacccharides and is obtained from red–purple seaweeds. The most well–known polysaccharides are starch, glycogen, cellulose and chitin.

Starch

Starch is a mixture of two glucans (polysaccharides built from glucose), which are called α–amylose and amylopectin. It is synthesized by plants as their principal food reserve and deposited in the plant cell cytoplasm as insoluble granules.

α–Amylose is a linear polymer of several thousand glucose residues, α–(1,4)–glycosidically linked. This polymer adopts an irregularly aggregating helically coiled conformation containing regular left–handed helix regions. Amylopectin on the other hand, carries α–(1,6)–connected branches every 24 to 30 glucose residues of the α–(1,4)–linked chain, resulting in a tree– or brush–like structure. It contains up to a million glucose residues which makes it among the largest molecules occurring in nature.

Starch is degraded by enzymes called amylases, which randomly hydrolyze the α–(1,4)–glycosidic bonds digesting the polysaccharide into oligosaccharide fragments, such as maltose and maltotriose as well as oligosaccharides containing α–(1,6)–branches, the latter being called dextrins. The oligosaccharides produced by amylase digestion are further hydrolyzed to glucose by specific glucosidases and by debranching enzymes, which remove the α–(1,6)–branches.

Glycogen

Glycogen is the storage polysaccharide of animals and is present in all cells but most pre-valently in skeletal muscle and liver, where it occurs in cytoplasmic granules. The primary structure of glycogen only differs from that of amylopectin in that it is more highly bran-ched with branching points occurring every 8 to 12 glucose residues of the α–(1,4)–linked glucan chain. The degree of polymerization of glycogen is similar to that of amylopectin. In the cell, glycogen is degraded for metabolic use by glycogen phosphorylase, which cleaves the α–(1,4)–linkages sequentially inwards from its non–reducing ends to release glucose–1–phosphate which can be fed into the citric acid cycle. The branching points are hydrolyzed by glycogen debranching enzyme. Glycogen contains about 1% covalently linked protein.

Glucose moieties, α–(1,4)–linked such as in starch and glycogen; these polymers are fur-ther modified by α–(1,6)–branches, which are not shown.

Glucose moieties, β–(1,4)–linked such as in cellulose. As a consequence of the β–linkages the three–dimensional shape and the proper-ties of cellulose are fundamentally different from those of starch and glycogen.

N–Acetylglucosamine residues, β–(1,4)–linked such as in chitin, which is similar to cellulose in shape and properties.

The disaccharidic unit of murein. This struc-ture differs from that of chitin only by an O–lactic acid group in the 3–position of every second GlcNAc residue.

Cellulose

Cellulose is an abundant carbohydrate of commercial and biological importance, found in all plants as the major structural component of the cell walls. Cellulose in wood is mixed with many other polymers such as hemicelluloses and lignin. It has to be split from these components to be used for paper production. The fluffy fiber found in the cotton ball is the purest naturally occurring form of cellulose.

Cellulose is the β–isomer of amylose consisting of β–(1,4)–linked glucose residues. The different stereochemistry of the glycosidic linkage compared to amylose gives cellulose totally different properties. In contrast to amylose, the β–linkages in cellulose allow the polymer to fold in a fully extended conformation to form a sheet–like secondary structure. The tertiary structure of cellulose is characterized by parallel–running intermolecular hydrogen–bonded cellulose chains further associated by hydrogen bonds and van der Waals forces to produce three–dimensional microfibrils. This gives cellulose fibres exceptional strength and makes them water insoluble despite their hydrophilicity. The cellulose microfibrils give an X–ray diffraction pattern that indicates regular, repeating microcrystalline structures interspersed by less–ordered paracrystalline regions.

As a consequence of its three–dimensional structure, cellulose cannot be hydrolyzed by starch–degrading enzymes. The cellulose–degrading enzymes, called cellulases, are produced by microorganisms.

Chitin

Chitin is the principal structural component of the exoskeleton of invertebrates such as crustaceans, insects, and spiders and is also present in the cell walls of most fungi and many algae, thus representing the widespread occurence of carbohydrates on earth, their abundance being almost equivalent to that of cellulose. The structure of chitin is related to that of cellulose. It is a homopolymer of β–(1,4)–linked N–acetyl–D–glucosamine residues. X–ray analysis indicates that chitin and cellulose have similar secondary and tertiary structures. De–N–acetylated chitin is called chitosan.

Murein

Murein, also named 'peptidoglycan', is an aminosugar polymer consisting of two β–(1,4)–linked hexoses, N-acetylglucosamin (GlcNAc) and N–acetyl–D–muraminic acid (NAMA), the latter differing from GlcNAc in that the 3–postion is substituted with an O–lactic acid group. The carbohydrate moieties are linked to short oligopeptides, which cross-link the polysaccharide chains. Murein is the structural component of the cell wall of all bacterial species. There are also other kinds of carbohydrates found in the bacterial cell wall, but it is murein that is the major unifying structure.

Pectins

Pectins are polysaccharides occurring in all plants primarily in their cell walls. They act as intracellular cementing material that gives body to fruits and helps them keep their shape. When fruit becomes overripe, the pectin is broken down into its monosaccharide constituents. As a result, the fruit becomes soft and loses its firmness. One of the most prominent characteristics of pectins is their ability to form gels. All pectins are composed of D–galactopyranosyl uronic acid units, which are α–(1,4)–linked. They contain methyl esters and acetyl groups to various degrees and show a typical average molecular size of 100,000 Da.

Dextrans

It has long been known that sucrose solutions convert into viscous solutions, gels, or flocculent precipitates. The material that produced such changes in sucrose solutions was isolated and found to be a polysaccharide that was called dextran. The enzymes capable of synthesizing polysaccharides from sucrose are produced by Gram–positive bacteria of the type *Leuconostoc* and *Streptococcus*.

A dextran is defined as a glucan that is enzymatically synthesized from sucrose and has contiguous α–(1,6)–linked D–glucopyranose units in the main chains. All known dextrans are branched at different branching points and to different extents. Differences in the three–dimensional structures of different dextrans are due to the percentage and manner in which the branches are arranged.

Glycosaminoglycans

The extracellular spaces, particularly those of connective tissues such as cartilage, tendon skin, and blood vessel walls, consist of collagen and elastin fibers embedded in a gel–like matrix known as ground substance. Ground substance is composed largely of glycosaminoglycans also called mucopolysaccharides because solutions of glycosaminoglycans have a slimy, mucus–like consistency that results from their high viscosity and elasticity. Glycosaminoglycans are unbranched polymers of alternating uronic acid and hexosamine residues. The most prevalent structures of these generally heterogeneous substances are hyaluronic acid, chondroitin–4–sulfate, chondroitin–6–sulfate, dermatan sulfate, keratan sulfate, heparin, and heparan sulfate.

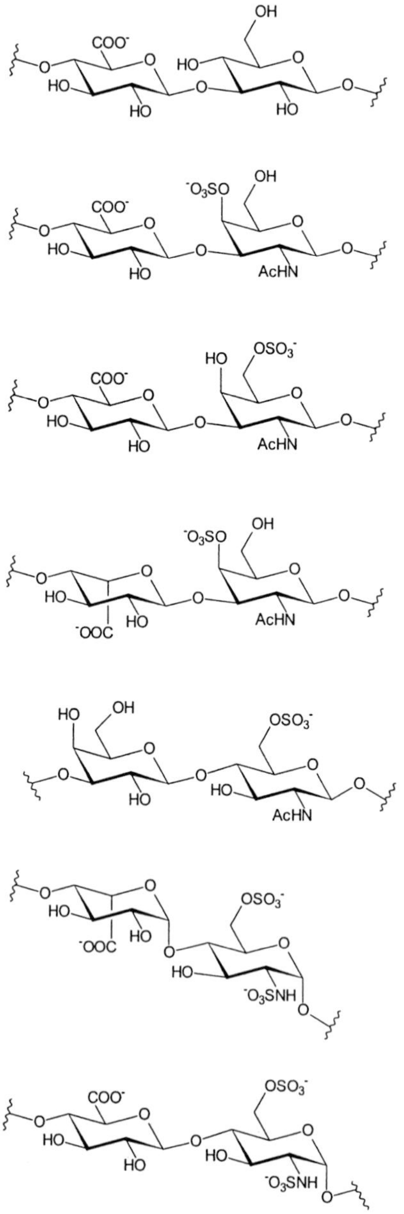

Hyaluronate

Chondroitin-4-sulfate

Chondroitin-6-sulfate

Dermatan sulfate

Keratan sulfate

Heparin

Heparan-sulfate

Disaccharide repeating units of the common glycosaminoglycans.

Hyaluronic acid is an important glycosaminoglycan of ground substance. It is widely distributed in mammalian cells and tissues and also in the capsules surrounding certain, usually pathogenic bacteria such as *Streptococci*. It consists of a linear chain of repeating disaccharide units of glucuronic acid and glucose and is composed of 500 to 50,000 monosaccharide units per molecule. Hyaluronate is a rigid, highly hydrated molecule. Its anionic character allows it to tightly bind cations such as K^+, Na^+, and Ca^{2+}.

The two chondroitin sulfates, containing *O*–sulfated GlcNAc as well as glucuronic acid, occur separately or in mixtures as major component of cartilage and other connective tissues. Dermatan sulfate, like heparin, contains L–iduronic acid. It is so–named because of its prevalence in skin.

Keratan sulfate is the most heterogeneous of the major glycosaminoglycans in that its sulfate content is variable and it also contains small amounts of fucose, mannose, *N*–acetyl–glucosamine and sialic acid.

Heparin is a variably sulfated glycosaminoglycan that consists predominantly of alternating α–(1,4)–linked residues of L–iduronic acid and *N*–, *O*–di–sulfated glucosamine. Heparin, in contrast to the above mentioned glycosaminoglycans is not a constituent of connective tissue but occurs almost exclusively in the intracellular granules of the mast cells that line arterial walls, especially in the liver, lungs, and skin. It inhibits clotting of blood and its release, through injury, prevents run–away clot formation. Therefore, heparin is in wide clinical use to inhibit blood clotting in postsurgical patients.

Heparin's anticoagulant effect has been attributed to its binding to antithrombin III, an important factor in the blood clotting cascade. A pentasaccharidic fragment has been identified as the minimal effective structural element in heparin binding to antithrombin based on the stereochemically favorable electrostatic interactions of the heparin sulfate groups and the protonated lysine–NH$_2$ groups of the protein.

Heparan sulfate occurs ubiquitously on cell surfaces but also occurs as an extracellular substance in blood vessel walls and brain. It resembles heparin but has a far more variable composition with fewer *N*– and *O*–sulfate groups and more *N*–acetyl groups.

The pentasaccharide segment in heparin which is responsible for binding to antithrombin III.

Proteoglycans

Proteins and glycosaminoglycans in ground substance aggregate covalently and non–covalently to form a diverse group of macromolecules known as proteoglycans, sometimes also called mucoproteins.

Proteoglycan subunits consist of a central core protein, which has three saccharide binding regions: The inner region, predominantly binds *N*–linked oligosaccharides, the central region binds *O*–linked oligosaccharides many of which bear keratan sulfate chains and the outer region, which binds chondroitin sulfate chains that are *O*–linked to the core protein via a galactose–galactose–xylose linker. The glycosylated core protein is non–covalently anchored to hyaluronic acid via its globular *N*–terminal end. This association is stabilized by a 'link protein'.

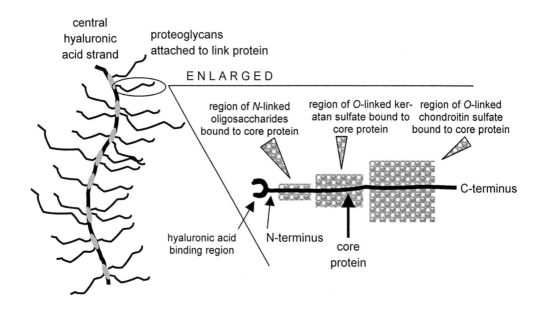

Cartoon to symbolize the structure of proteoglycans and their non–covalent complex formation with hyaluronic acid.

Structure analysis of polysaccharides

The analysis of polysaccharide structures is of general interest but also of industrial relevance as the structure and properties of polysaccharides are closely related. Polysaccharide analysis requires specialized techniques, which differ from those methods used for the characterization of small molecules.

For structure analysis of polysaccharides the following aspects have to be elucidated:

(i)   nature and molar ratios of the contained monosaccharide building blocks;
(ii)   positions of the glycosidic linkages;
(iii)   distinction of furanosidic and pyranosidic forms;
(iv)   anomeric configuration;
(v)   monomer sequences and identification of repeating units; and
(vi)   position and nature of OH–modifications such as *O*–phosphorylation or *O*–sulfation.

The challenges listed above, which do not even include supramolecular considerations, show that the structural analysis of polysaccharides as well as of complex oligosaccharides, can be a complex and demanding task. Mass spectrometry can provide much information regarding sequence analysis and the identification of building blocks. However, for the analysis of the individual monomer types, the polysaccharide is normally subjected to a total hydrolysis under strongly acidic conditions, followed by reduction and peracetylation of the resulting monomeric units. The acetylated alditols that are formed are then subjected to chromatographic analysis and their retention times recorded.

Elucidation of linkage positions is achieved by permethylation of the polysaccharide. Acidic hydrolysis of the resulting poly–methylethers cleaves only the interglycosidic linkages and leaves the methylether bonds intact. Reduction and acetylation then yields partially methylated alditols, which are acetylated at the former linkage positions. The products of this so–called standard methylation analysis are then characterized by gas chromatography and mass spectrometry.

Standard methylation analysis however, has one major drawback and that is that during the reduction step leading to the alditols, structural information is lost, for example by the formation of *meso*–alditols. Furthermore, the same alditol structure is formed in this procedure regardless of whether a 4–*O*–linked aldopyranose or the corresponding 5–*O*–linked aldofuranose was present in the polysaccharide. A solution to this problem was found by using a reductive cleavage protocol, during which the methylated polysaccharide is cleaved using a Lewis acid instead rather than a Brönsted acid; this is then followed by acetylation of the fragments obtained. This methodology was introduced by G. Gray. It yields partially methylated anhydro alditols in which the anomeric position is deoxygenated and only the second linkage position of the sugar ring is acetylated. Thus different reaction products are obtained depending on whether furanosides or pyranosides were cleaved. A combination of trimethylsilymethanesulfonic acid (TMS–mesylate) and $BF_3 \cdot Et_2O$ proved to be very successful in this procedure.

Standard methylation analysis

Reductive cleavage method

By standard methylation analysis of polysaccharides less structural information is obtained than by employing the reductive cleavage method. While standard analysis does not allow the two residues C and D to be distinguished, when they have the same relative configurations at the stereogenic centers, their different ring forms (pyranose and furanose form, respectively) are conserved in the reductive cleavage method.

3   Protecting groups for carbohydrates

Regioselectivity is a prominent problem in carbohydrate chemistry as sugars contain several hydroxyl groups of different reactivity together with various other functionalities such as carbonyl and amino groups. Protecting groups and protecting group strategies are, therefore, of crucial importance for carbohydrate chemistry. For the synthesis of special mono- or oligosaccharide derivatives or glycoconjugates as well as for the preparation of selectively protected glycosyl acceptors for glycoside synthesis, an orthogonal set of protecting groups has to be chosen which allows the broadest possible flexibility for the planned synthetic avenues by which each protecting group can be split off without affecting the rest of the molecule. With modern protecting groups there is the potential of fulfilling every possible protection pattern. However, a good protecting group strategy remains a central challenge in sugar chemistry as certain protecting groups may also influence the reactivity of carbohydrate derivatives such as glycosyl donors and glycosyl acceptors, for steric as well as electronic reasons. Protecting groups, especially those adjacent to the anomeric center will also influence the stereoselectivity of glycosidation reactions and can be chosen as neighboring active or inactive groups.

In carbohydrates hydroxyl groups have to be protected most frequently. They differ in their reactivity depending on whether they are anomeric, primary or secondary hydroxyls. These differences in reactivity can sometimes be utilized so that a desired protection pattern can be achieved in one step without using a more complex reaction sequence.

The most common protecting groups for hydroxyl functions are esters, ethers, and acetals. They will be discussed in this chapter as well as methods for the protection of amino groups in 2–amino sugars.

3.1   Protection of the anomeric center

Due to the special reactivity of the anomeric hydroxyl group in carbohydrates, this is usually protected prior to all other OH–functionalities. Such anomerically protected derivatives are mostly simple glycosides, which can be prepared by Fischer glycosylation. Fischer glycosylation involves the acid–catalyzed reaction of the cyclic hemiacetal (the free sugar) with an alcohol to form the respective acetal, called a glycoside. The equilibrium for this reaction has to be shifted towards the product by using an excess of alcohol, because water is a by-product of the reaction and causes hydrolysis (retroreaction) of the formed acetal under the acidic conditions applied. As the excess alcohol has to be distilled off after conversion of the sugar is completed, Fischer glycosylation is limited to the use of low–boiling alcohols such as methanol, ethanol, and allyl alcohol, respectively, and sometimes even benzyl alcohol is glycosidated in a Fischer reaction. As an acid catalyst acidic ion exchange resin is the most convenient to use, as it can be removed by simple filtration once the reaction is completed. Fischer glycosylations yield a mixture of stereoisomeric glycosides, of which the pyranoside, which is favored by the anomeric effect, is normally obtained as the main product and can often be crystallized from the mixture.

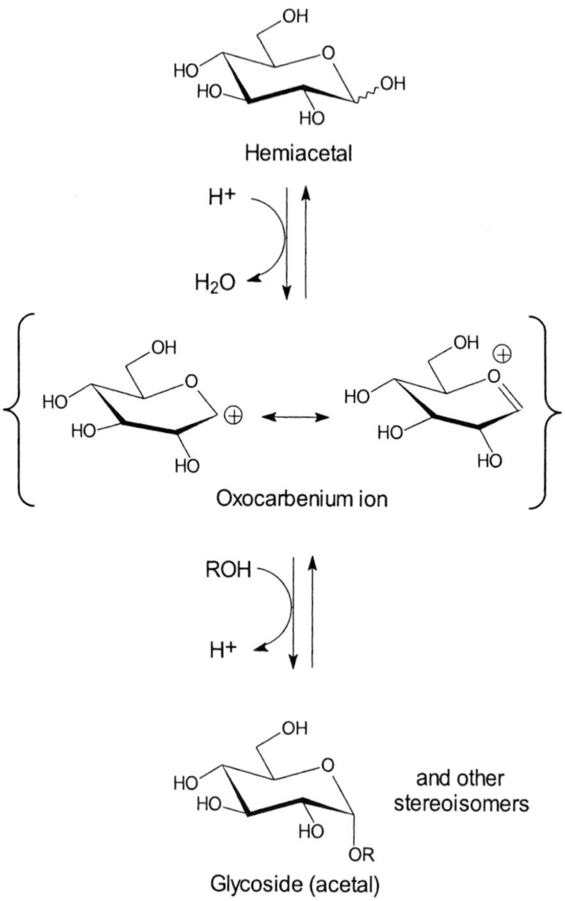

Hemiacetal

Oxocarbenium ion

Glycoside (acetal)

and other stereoisomers

Mechanism of the Fischer glycosylation of alcohols (ROH). The reducing sugar (a hemiacetal) is thereby converted to a glycoside (an acetal).

Synthesis of methyl α–L–fucoside by Fischer glycosylation.[1]

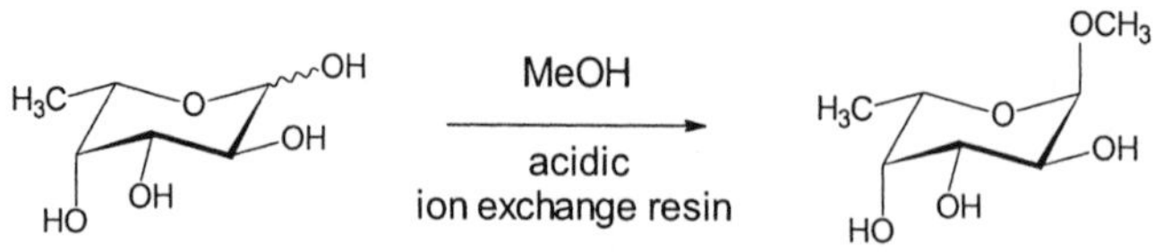

10 g (60.9 mmol) L–fucose are dissolved in 100 ml of dry methanol and 10 g (wet weight) of strongly acidic ion exchange resin, which has been thoroughly washed with dry methanol, is added. This reaction mixture is heated under reflux for 18 h possibly without stirring so not to destroy the resin. One main product is identified by TLC (*n*–PrOH–EtOH–H₂O, 5:3:2).

The resin is filtered off, washed with methanol, and the filtrate concentrated *in vacuo*. The residual solid is recrystallized from ethanol to yield the desired methyl glycoside as colorless needles (9.1 g, 84%).

In order to be of use as anomeric building blocks, glycosides should be stable against a multitude of reagents, but should yet be amenable to selective removal or transformation of the anomeric group by reactions that are mild enough not to affect other functional groups in the molecule. This is a special problem with methyl glycosides which are among the most stable glycosides and require forced acid–catalyzed reaction conditions for cleavage.

Synthesis of 2,3,4,6–tetra–O–benzyl–D–mannopyranose by cleavage of methyl 2,3,4,6–tetra–O–benzyl–α–D–mannopranoside.[2]

In a round–bottomed flask, equipped with a condenser, 7.5 g (13.8 mmol) methyl 2,3,4,6–tetra–O–benzyl–α–D–mannopyranoside are dissolved in 38 ml acetic acid and 12 ml 3 M H_2SO_4 (2 ml conc. H_2SO_4 + 10 ml H_2O) and stirred at 85°C for 4.5 h. Then 50 ml of cold H_2O and 50 ml of toluene are added, the phases are separated and the aqueous phase is extracted three times with 75 ml of toluene each. The combined organic phases are washed with satd. aq. NaCl solution, dried over $MgSO_4$, filtered and evaporated *in vacuo*. The residue is purified by flash chromatography (toluene–ethyl acetate, 7:1) to yield the reducing sugar (3.5 g, 5.8 mmol, 42%) as a colorless syrup.

Alternatively methyl glycosides can be converted to glycosyl halides by reaction with dihalomethylmethyl ether and a Lewis acid under somewhat elevated temperatures using neat dichloromethylmethyl ether or additional solvent.

R = Me, Et, Pr, *i* Pr, *t* Bu, Bn, Ph

Other glycosides, which can also be obtained by Fischer glycosylation but are more easily cleaved than methyl glycosides are benzyl and allyl glycosides, respectively. Allyl glycosides can be cleaved by isomerization of the allyl to a 2–propenyl group, which is easily removed under mild acidic conditions. This isomerization can be effected with Pd (0) catalysts, for example (cf. chapter 3.3). Benzyl glycosides can be cleaved by catalytic hydrogenation.

R = **Me** : strong acidic conditions, high temperature

R = **Allyl** : [Pd], *p*-TsOH, elevated temperature
R = **Benzyl** : [Pd], H_2, room temperature

Synthesis of 2,3,4,6–tetra–O–benzyl–α–D–mannopyranose, a three–step synthesis.[3]

Step (i): [H+], reflux

Step (ii): BnCl, aq. NaOH, TBABr

Step (iii): [Pd], p-TsOH, MeOH, H₂O, reflux

Step (i): Synthesis of allyl α–D–mannoside by Fischer glycosylation. Portions of acetyl chloride (25 ml, 0.35 mol) are slowly added to 300 ml of allyl alcohol at 0°C, then D–mannose (25 g, 138.9 mmol) is added and the reaction mixture is stirred at 70°C for 4 h. The mixture is neutralized with solid NaHCO₃, filtered over a celite bed, and evaporated in vacuo. Repeated co–evaporation with toluene and flash-chromatographic purification (ethyl acetate-methanol, 17:3) gave 14.5 g (65.9 mmol, 48%) of the title mannoside as a slightly yellow syrup.

Step (ii): Synthesis of allyl 2,3,4,6–tetra–O–benzyl–α–D–mannopyranoside. Crude allyl α–D–mannoside (25 g) is dissolved at 55°C in 300 ml of 33% aq. NaOH solution. Tetrabutylammonium bromide (TBABr, 32 g) is added and benzyl chloride (77 ml) is added dropwise over 1 h. The reaction mixture is stirred at 55°C for 4.5 h and then at room temperature overnight. 200 ml of toluene are added, the phases are separated and the organic phase is washed with water until it is neutral. It is concentrated in vacuo and purified by flash chromatography (toluene–ethyl acetate, 35:1) to yield 36 g (63 mmol, 75%) of the fully protected product as colorless syrup.

Step (iii): Synthesis of 2,3,4,6–tetra–O–benzyl–α–D–mannopyranose. Allyl 2,3,4,6–tetra–O–benzyl–α–D–mannopyranoside (7.4 g, 12.7 mmol) is dissolved in 60 ml 85% aq. methanol and palladium on charcoal (Pd–C, 10%, 1.6 g) and p–toluenesulfonic acid (1.134 g, 5.96 mmol) are added to the solution. The reaction mixture is heated under reflux for 3 h. After cooling to room temperature, the mixture is filtered over a celite bed, concentrated in vacuo and then dissolved in 100 ml of toluene and dried over MgSO₄. Filtration, concentration and purification by flash chromatography (toluene–ethyl acetate, 7:1) gives the reducing mannose derivative (4.5 g, 8.3 mmol) in 65% yield as colorless syrup.

When oligosaccharide derivatives, rather than those of monosaccharides have to be temporarily protected at the anomeric center, removal of the anomeric protecting group must be achieved with reagents which cleave the aglycone but are mild enough not to break the intersaccharidic glycosidic bonds. For this purpose 2–(trimethyl–silyl)ethyl (TMSET) glycosides were introduced to carbohydrate chemistry by G. Magnusson.[4] They are easily formed employing Koenigs–Knorr glycosylation for example, (cf. section 4.3) using glycosyl halides and 2–(trimethylsilyl)ethanol. On the one hand, 2–(trimethylsilyl)ethyl glycosides are compatible with most reaction conditions used in carbohydrate synthesis and on the other hand, they can be converted to the hemiacetals, to glycosyl halides and the corresponding 1–*O*–acyl derivatives, respectively, under mild conditions, leaving intact other interglycosidic linkages of an oligosaccharide. This makes the anomeric 2–(trimethylsilyl)protecting group very useful.

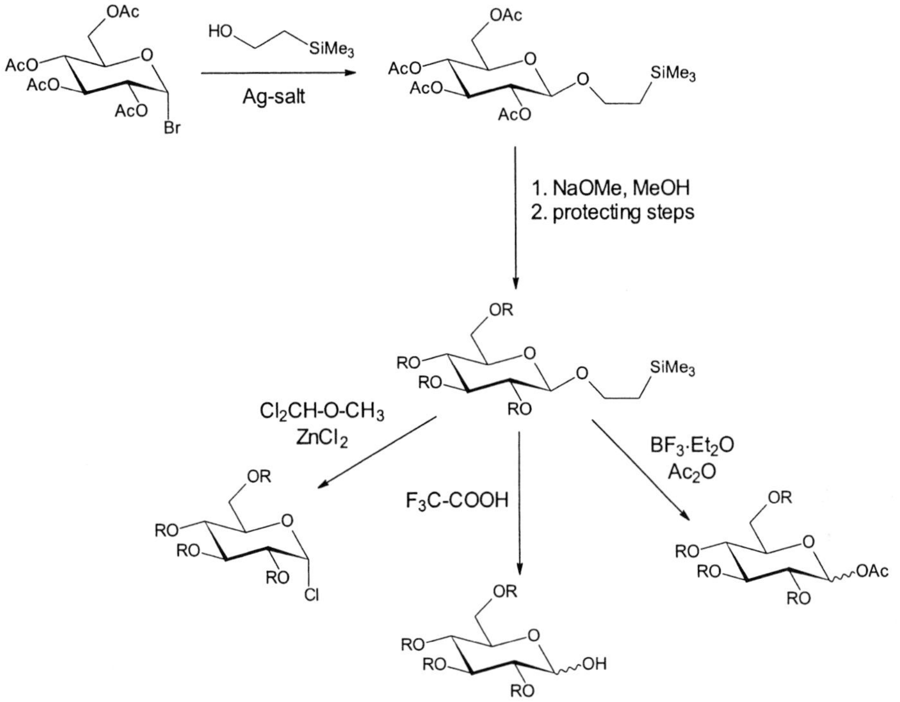

The increased reactivity of 2–(trimethylsilyl)ethyl glycosides is based on acid–induced fragmentation of the β–silicon–substituted organosilicon aglycone. Thus, the reducing oli-

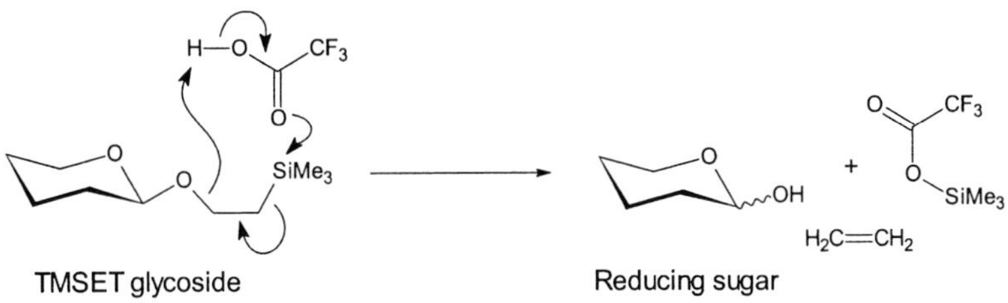

TMSET glycoside   Reducing sugar

gosaccharide can be obtained from its TMSET glycoside by reaction with trifluoroacetic acid in dichloromethane at room temperature. Nucleophilic attack at the silicon atom, leads to the release of ethylene and protonation of the anomeric oxy–anion to afford the hemiacetal.

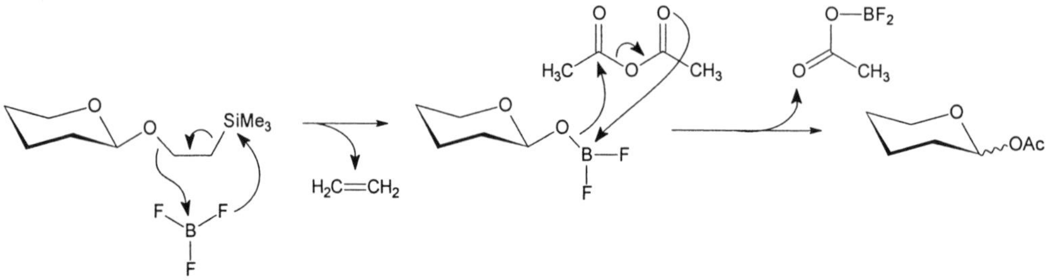

The analogous fragmentation reaction effected by boron trifluoride and acetic anhydride gives rise to the corresponding 1–*O*–acyl derivative and conversion of TMSET glycosides to glycosyl chlorides can be achieved with dichloromethylmethyl ether and zinc chloride as the Lewis acid.

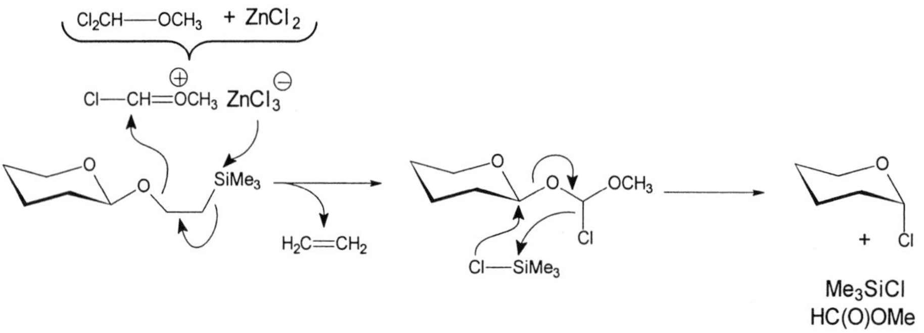

Thioglycosides and *n*–pentenyl glycosides can also be regarded as anomeric building blocks, as they can be converted to a number of anomeric derivatives or used as glycosyl donors after activation (cf. sections 4.5 and 4.6). In addition, 1,2–glycosyl orthoesters and 1,6–anhydro sugars, like levoglucosan, can be employed in synthetic pathways, which require temporary protection of the anomeric center. In the case of 1,6–anhydro sugars, acetolysis affords the 1,6–di–*O*–acetylated compounds; 1,2–glycosyl orthoesters are cleaved under acidic conditions to give 2–*O*–acetyl–1–OH derivatives.

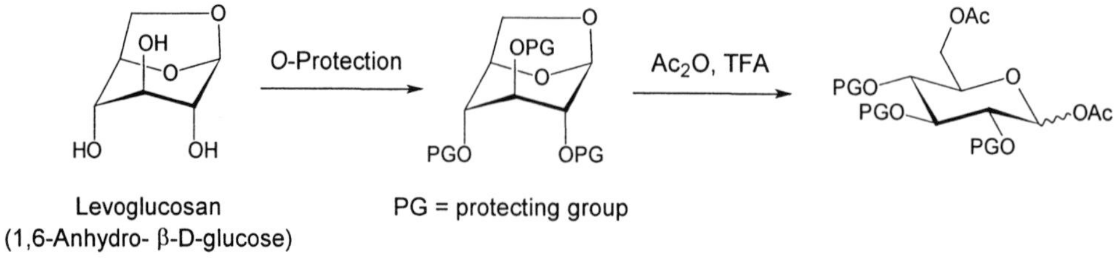

Levoglucosan
(1,6-Anhydro- β-D-glucose)

PG = protecting group

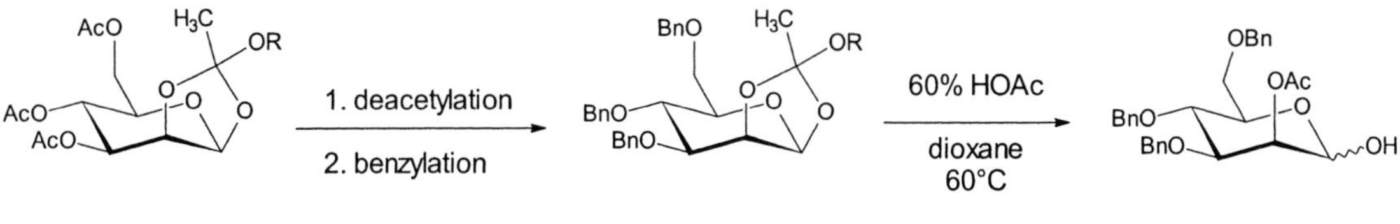

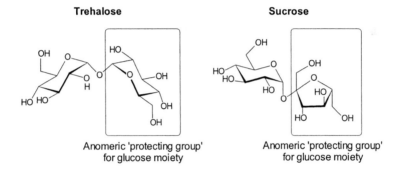

It can be a useful alternative for anomeric protection to employ non–reducing disaccharides, such as trehalose and sucrose. In these cases, one monomer part serves as an anomeric protecting group for the other. The interglycosidic linkage can eventually be cleaved selectively by chemical or enzymatic means, after the appropriate protection steps.

Trehalose

Sucrose

Anomeric 'protecting group'
for glucose moiety

Anomeric 'protecting group'
for glucose moiety

Protection of carbohydrates does not, however, always start with the selective protection of the anomeric center. It often involves peracylation as the first step leading to uniformly protected monosaccharides, which are suitable starting materials for a variety of conversions.

3.2   Acyl protecting groups

Acyl protecting groups are frequently used in carbohydrate chemistry and are easily introduced. Thus, the complete acetylation of a free sugar using acetic anhydride is often the first
step in a sequence of conversions. Alternatively perbenzoylations using benzoyl chloride as
the reagent may form the start of a synthetic sequence with an unprotected saccharide.
Complete acylations of carbohydrates lead to an anomeric mixture of the peracylated sugars, of which one anomer can often be selectively crystallized. Mostly, however, they are
further converted as anomeric mixtures. Fully acylated sugars are regularly used for the
preparation of glycosyl donors such as glycosyl bromides, 1–thio–glycosides, glycosyl trichloroacetimidates or are used themselves as glycosyl donors for the preparation of simpler
glycosides in a Lewis acid–catalyzed reaction (cf. chapter 4).

Esters are easily cleaved in basic medium. Instead of using saponification conditions, a
transesterification reaction is utilized, known as the Zemplén procedure. In this reaction
catalytic amounts of sodium methanolate in dry methanol are used for the deprotection of
the ester groups. This reaction may be completed after 10 minutes, but sometimes requires
longer reaction times. In general, benzoates are less easily cleaved than acetates.

Unprotected sugars, mono- or oligosaccharides can be fully acetylated (R=Ac) with acetic
anhydride or perbenzoylated (R=Bz) employing benzoyl chloride as the reagent. Deprotection of these derivatives succeeds in an easy and quick transesterification reaction, called
the Zemplén procedure, using sodium methanolate in methanol.

For the peracetylation of multigram amounts of unprotected mono- and disaccharides the
sugars are heated in acetic anhydride, adding sodium acetate as buffer. Under these conditions, a larger amount of the β–anomer is formed. For more sensitive sugars and especially
for the acetylation of only one or a few hydroxyl groups in a functionalized carbohydrate
derivative, the acetylation with acetic anhydride is carried out in pyridine at room temperature. In this case, N–acetylpyridinium acetate is formed first and acts as the actual acetylat-
ing species. Benzoylations are normally carried out with benzoyl chloride in pyridine at
room temperature. Acylations in pyridine can be accelerated by the addition of DMAP to
the reaction mixture.

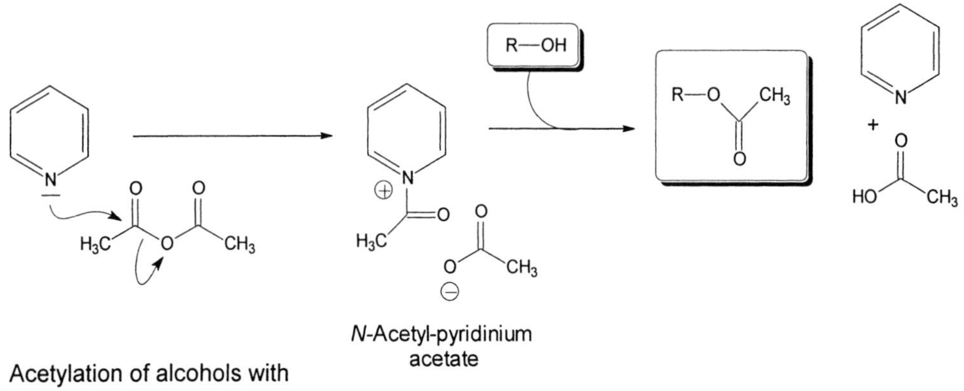

Acetylation of alcohols with acetic anhydride in pyridine.

N-Acetyl-pyridinium acetate

Synthesis of penta–*O*–acetyl–β–D–glucopyranose.[5]

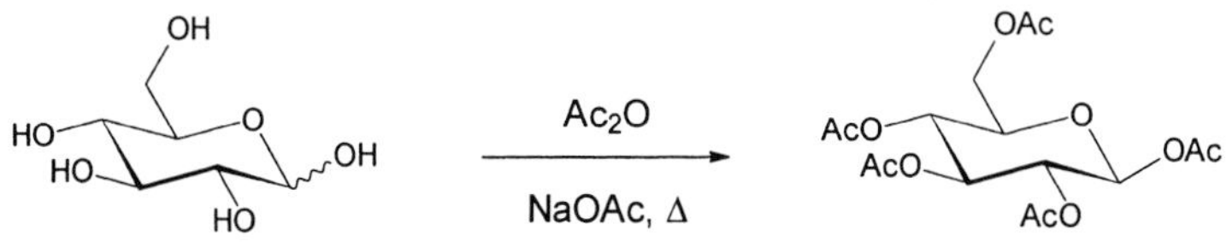

A suspension of 25 g (0.3 mmol) anhydrous sodium acetate in 300 ml (3.2 mmol) of acetic anhydride is heated to reflux temperature in a round–bottomed flask. Then, the heating is removed and anhydrous D–glucose (50 g, 0.3 mmol) is added in portions, so that the reaction mixture continues to reflux. When all glucose is added, the reaction mixture is stirred under reflux for 30 min, poured onto 1 liter of crushed ice and stirred for 2 h. The precipitated product can then be filtered off, washed with water and recrystallized from ethanol to yield 53 g (50%) colorless crystals melting at 135°C.

General acetylation procedure using acetic anhydride–pyridine. The compound to be acetylated is dissolved in dry pyridine (approx. 10 ml pyridine for 1 g of sugar), acetic anhydride (2 equivalents per OH–group) is added and the mixture is stirred at room temperature until the reaction is complete (observed using TLC). In the case of small scale re-actions (less than approx. 500 mg starting material) for work–up, co–evaporation with toluene followed by flash chromatography is sufficient to obtain pure acetylated product. When larger amounts of acetic anhydride are used, an aqueous work–up is advisable prior to flash chromatography.

General benzoylation procedure. The compound to be benzoylated is dissolved in dry pyridine (approx. 10 ml pyridine for 1 g of sugar), benzoyl chloride (2 equivalents per OH–group) is added and the mixture is stirred at room temperature until the reaction is complete (observed using TLC). Water is added, and the solution is extracted three times with di-chloromethane. The combined organic phases are successively washed with satd. aq. Na-

HCO_3 solution, 1N HCl, and water, dried over $MgSO_4$, filtered and concentrated *in vacuo*. Flash chromatography yields the pure benzoylated product.

General deacylation procedure according to Zemplén.[6] The compound to be de-acetylated is dissolved in dry methanol (approx. 10 ml methanol for 1 g of sugar), approx. 100 µl of a freshly–prepared 1M sodium methanolate solution is added and the mixture is stirred at room temperature until the reaction is complete (observed using TLC). Then acidic ion exchange resin is added for neutralization, the mixture is stirred for a few minutes and after neutralization (monitored using pH paper) the resin is filtered off, washed with metha-nol and the filtrate concentrated *in vacuo*. The product is normally pure, but, if necessary, can be purified with flash chromatography.

Acetyl and benzoyl groups are the common esters for protection, however, the arsenal of acyl protecting groups includes a number of other useful members. The sterically demand-ing pivaloyl (2,2–dimethylpropanoyl) protecting groups are employed when acyl migration has to be avoided or when carbohydrate derivatives are employed as chiral auxilliaries. Pi-valoates are less prone to acyl group migrations and have a much lower tendency to give orthoesters in Koenigs–Knorr glycosidations. They are cleaved by aqueous methylamine at room temperature. They can be distinguished from acetyl protecting groups as they are not cleaved by aqueous ammonia, which removes acetates. Moreover, pivaloates can be or-thogonally used together with phthalimide protected amino groups, because they are not cleaved by hydrazine in ethanol even at reflux, whereas phthalimides are removed under these conditions (cf. section 3.6).

Chloroacetyl and bromoacetyl protecting groups can be used orthogonally to acetyl groups because they can be cleaved by thiourea under essentially neutral conditions which leave acetyl and benzoyl groups intact. Thiourea, however, often gives rise to acyl group migration and, therefore, hydrazine is often used instead and it is also the reagent of choice when more than one chloroacetyl group has to be removed.

Benzoyl groups are less easily cleaved than acetyl groups and the two can therefore be discriminated by certain deprotection procedures. DBU in dichloromethane, for example, removes most acetate protecting groups and leaves benzoates intact.

Regioselective acylation

Acylations can often be carried out as regioselective reactions to establish selective protection patterns. Regioselective benzoylations are more feasible than selective acetylations. Typically equatorial hydroxyl groups can be benzoylated prior to axial hydroxyls at lowered temperatures. This technique is well utilized in the *galacto*–series, for example in the selective benzoylation of methyl α–D–galactoside, affording the 2,3,6–tri–O–benzoylated derivative in yields up to 90%.

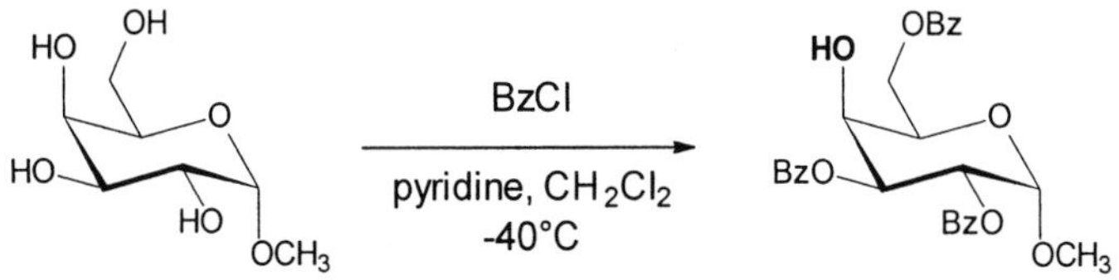

By analogy, L–fucose (6–deoxy–L–galactose) can be selectively benzoylated to yield 1,2,3–tri–O–α–L–fucose in yields around 70%. The order of reactivity of the hydroxyl groups in a pyranose ring is different for every monosaccharide. In fucose it has been found to be 1–OH > 3–OH > 2–OH > 4–OH.

Synthesis of 1,2,3–tri–O–benzoyl–6–deoxy–α–L–galactopyranose.[7]

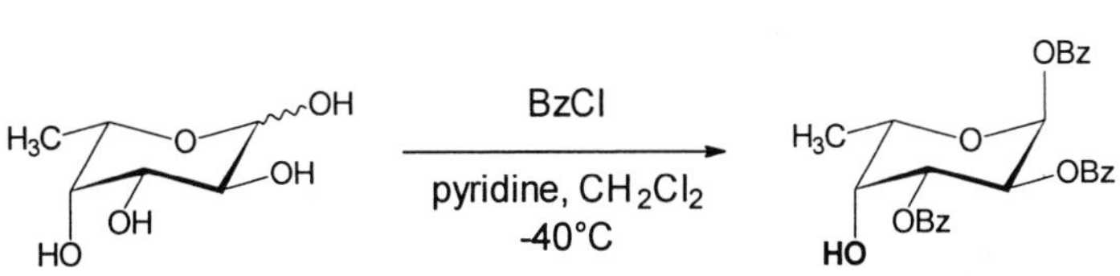

To a solution of 1 g L–fucose (6–deoxy–α–L–galactopyranose) in 10 ml dry pyridine a solution of 2.3 ml benzoyl chloride in 10 ml dry pyridine is added dropwise at -40°C. The mixture is stirred at -40°C for about 15 minutes and then quenched with the addition of water. The solution is extracted with dichloromethane several times, and the combined organic phases are co–evaporated with toluene. 1.9 g (65%) of the desired product is obtained after chromatographic purification (toluene–ethyl acetate, 4:1) as a white foam.

In addition, regioselective acylation reactions can be performed using additives such as 1–hydroxy–1*H*–benzotriazole. In these procedures, the acylating species, 1–acyloxy–1*H*–benzotriazole can be prepared *in situ* from the triazole and the desired acyl chloride.

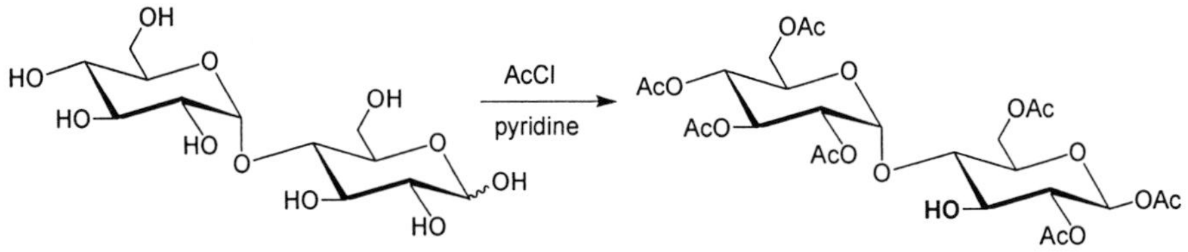

Most of the chemistry used for monosaccharides can normally equally be applied to disaccharides. Sometimes advantage can be taken from the different steric environment of homologous regions of a disaccharide. Thus maltose can be selectively modified at the 6'–postion (by tritylation or tosylation, for example) due to its lower steric hindrance compared with the 6–position. Moreover, acetylation or benzoylation of maltose carried out at room temperature leaves the 3–hydroxyl group, which is the least reactive OH in the disaccharide, free due to steric hindrance, as shown.

Selective deacylation at the anomeric center

Selective deacylation at the anomeric center of mono- and oligosaccharides can be carried out in high yields with a variety of reagents. The resulting reducing sugars are important intermediates in many syntheses, especially in the preparation of glycosyl trichloroacetimidates, which serve as glycosyl donors in glycosylation reactions (cf. section 4.4).

Using one equivalent of hydrazine acetate or hydrazine hydrate in DMF is the classical procedure for regioselective anomeric deacetylation and works very well. Piperidine and 2–aminoethanol have been employed as alternatives. Recently, ethylenediamine in a mixture with acetic acid has been found useful to selectively effect regioselective anomeric deacylation of fully acylated carbohydrates in yields over 90%. For anomeric debenzoylation ethanolic dimethylamine in pyridine gives good results. In general, anomeric debenzoylations take longer than the respective deacetylations.

General procedure for anomeric deacetylation using acetic acid–ethylenediamine.[8] To a solution of ethylenediamine (1.2 mmol) in THF (25 ml), glacial acetic acid (1.4 mmol) is added dropwise resulting in immediate formation of a precipitate which remains present until aqueous work–up. The starting peracetate (1 mmol) is added and the mixture is stirred 16–24 h. TLC (hexane–acetone, 3:2) then shows the absence of the starting material and the presence of a slower–moving product, which appears mostly as an anomeric mixture. Water (10 ml) is added and the mixture is extracted with dichloromethane. The organic phase is subsequently washed with 2N HCl, satd. aq. NaHCO$_3$ solution and water, dried over MgSO$_4$ and concentrated *in vacuo*. Purification by flash chromatography, sometimes followed by crystallization yields the desired products in yields over 90%.

General procedure for anomeric debenzoylation using pyridine–ethanolic dimethylamine.[9] The fully benzoylated sugar (5 g) is completely dissolved in 50 ml of dry pyridine and 35 ml of an ethanolic solution of dimethylamine (5.6 M) is added. The reaction mixture is stirred at room temperature and monitored by TLC every 10 minutes (toluene–ethyl acetate, 5:1). The reaction is quenched by addition of 100 ml toluene, when a slower moving product under the main product starts to show up clearly, indicating cleavage of further benzoyl groups in addition to the anomeric benzoyl group. Reaction times vary between 1h and 3 h. The mixture is washed three times with satd. aq. NaCl solution, and the organic phase is dried over MgSO$_4$, filtered, concentrated *in vacuo*, and purified by column chromatography (toluene–ethyl acetate, 5:1). This procedure can also be used for the anomeric deacetylation under slightly varied, milder reaction conditions.

Migration of acyl groups

Acyl groups have a high tendency to change their position at the carbohydrate ring, and this migration process can be base–, and also acid–catalyzed. Acetyl groups are most prone to migrations, whereas benzoyl groups migrate less easily, and by using pivaloates, acyl migration can usually be totally avoided. Acetyl group migrations are found to occur as intra- as well as intermolecular processes. Intermolecular acetyl migration has been observed during glycosylation of unreactive alcohols, diminishing the yield of the reaction significantly.

Intermolecular
acetyl migration

Certain characteristics of the saccharide favor intramolecular acetyl group migrations. For instance, when the anomeric center of a carbohydrate derivative is unprotected, acetyl migrations toward the anomeric center are observed. In the case of a non–reducing saccharide, migrations frequently occur 'down the chain', say toward the 6–position of the sugar ring. Frequently O→N acyl group migrations are observed. This is a severe problem in many syntheses where an azide group has to be reduced to yield an amino function. The catalytic hydrogenation reaction in these cases is often accompanied by acetylation of the *in situ* generated amino group caused by acetyl migration from a hydroxyl. This is an undesired side reaction in the majority of cases, as *N*–actylated amino groups can only be deprotected under harsh basic conditions.

O→N Acyl group migrations have been avoided in the case of reduction of the azido group of (2–azioethyl) 2,3,4,6–tetra–*O*–acetyl–β–D–glucopyranoside by adding Boc anhydride (di–*t*–butyldicarbonate) to the reaction mixture. In this way the amino group is immediately protected by Boc anhydride once it is formed from the azido group.

Synthesis of (2–*t*–butyloxycarbonylamidoethyl) 2,3,4,6–tetra–O–acetyl–β–D–gluco–pyranoside.[10]

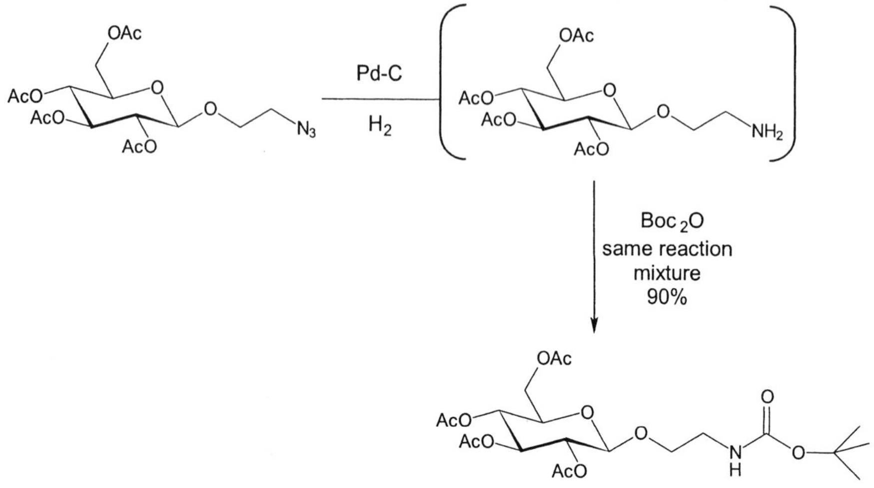

A suspension of (2–azidoethyl)–2,3,4,6–tetra–O–acetyl–β–D–glucopyranoside (10.5 g, 0.025 mmol), di–*t*–butyldicarbonate (8.18 g , 3.75 mmol) and activated palladium on charcoal (10%, 0.1 g) in ethyl acetate is stirred under a hydrogen atmosphere (1 bar) for 6 h at room temperature. The reaction mixture is filtered through a celite bed, the filtrate is washed with water and brine and the organic phase dried over anhydrous sodium sulfate. It is filtered, evaporated on a rotary evaporator and the crude product is purified by column chromatography (ethyl acetate–petroleum ether, 1:2) to yield the Boc protected amine (11 g, 90%) as a colorless syrup.

On the other hand, acyl group migrations have been used deliberately to establish protection patterns which are otherwise more difficult to obtain. Closely related to acyl group migration is the rearrangement of acyloxonium salts. The hexachloroantimonates of these can be obtained from acetylated glycosyl chlorides with antimony pentachloride in dichloromethane. The acetoxonium ion which is so obtained from 2,3,4,6–tetra–O–acetyl–β–D–glucopyranosyl chloride undergoes a series of acetoxonium ion rearrangements as shown on the next page, in which the configuration at the carbon centers which are involved in the rearrangement, change. Because the *ido*–configured acetoxonium ion is the least soluble in dichloromethane it is only this configuration that crystallizes out of the solution. This procedure allows to prepare the rare monosaccharide D–idose from D–glucose following the hydrolysis of the *ido*–configured acetoxonium ion. The procedure was discovered by H. Paulsen and co–workers.[11]

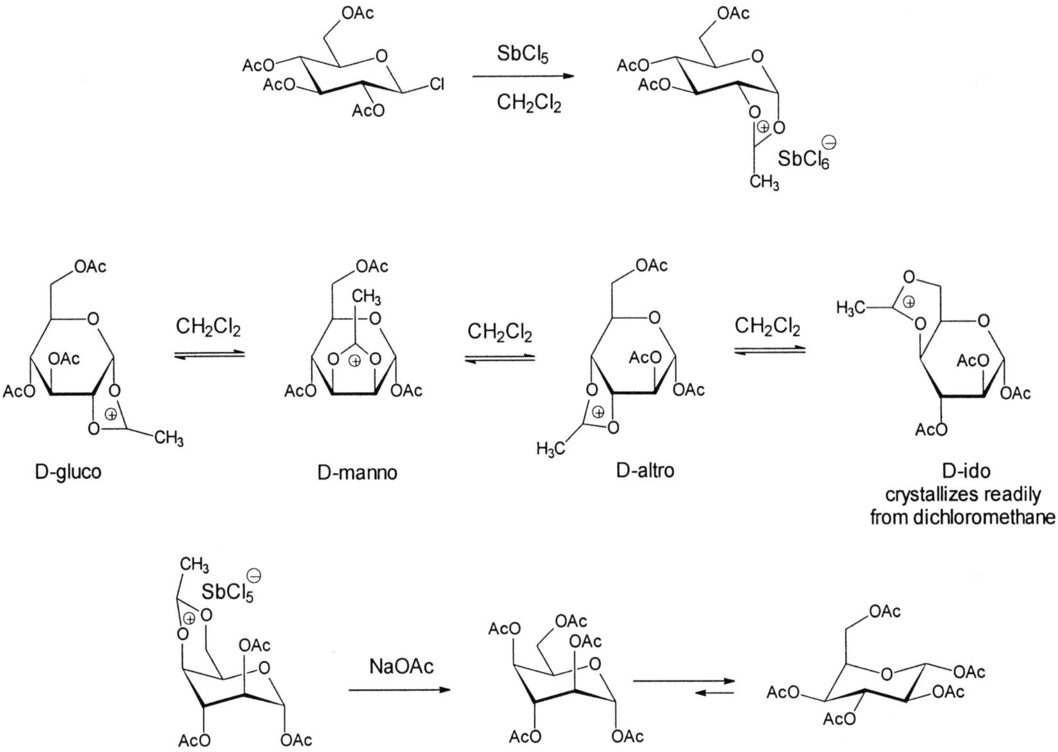

3.3   Ether protecting groups

Alkyl ethers are particularly stable entities, resistant to strong base and also to the acidic conditions which can be tolerated in glycoside chemistry. Methyl ethers, for example, are therefore only used for polysaccharide analysis and not as protecting groups. For the temporary protection of carbohydrates only those special ethers can be employed, which are cleavable under appropriate reaction conditions. Consequently, benzyl ethers, which can be removed by catalytic hydrogenation, and allyl ethers, unstable after isomerization into the corresponding vinyl ether, are the frequently used ether protecting groups in carbohydrate chemistry. Together with ester groups they already form a potent set of protecting groups resulting in a great variety of orthogonal protection within a saccharide.

Ethers are formed under the classical conditions of a Williamson synthesis, employing sodium hydride or sodium hydroxide as the base in a polar aprotic solvent such as DMF together with the respective alkyl bromide or alkyl chloride. Other than in the case of esterification reactions, it is not feasible to convert reducing saccharides under Williamson conditions, as they are barely stable under strongly basic conditions. Also base labile protecting groups such as acetates do not survive the etherification reactions. Therefore, it is not possible to plan a protection sequence so that esterification is conducted first and then followed by etherification. Ether protection should always be carried out before esterification, when both protecting groups are used in one saccharide. Ether protection also implies, that the saccharides in this way protected become more reactive in glycosylation reactions than their more electron deficient acylated counterparts, both as glycosyl donors as well as glycosyl acceptor.

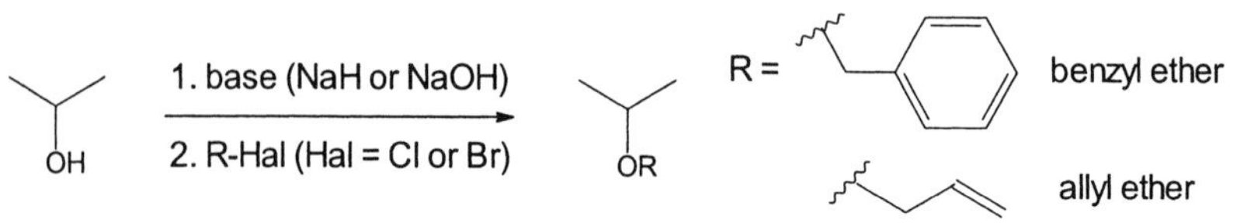

Benzyl ethers

For catalytic hydrogenation of benzyl ethers most frequently palladium on charcoal (Pd–C, 10%) is used as the catalyst. This is a quantitative reaction in the majority of cases. But occasionally debenzylation is a difficult step and then high hydrogen pressure of up to 100 bar is required in the hydrogenation reaction.

Substituted benzyl ethers are sometimes used in carbohydrate chemistry to expand the orthogonal possibilities of a protecting group set. The *p*–methoxybenzyl (*p*MBn) group is the most widely utilized benzyl ether alternative in this regard. It is introduced by the same methods as benzyl ethers, but it can be cleaved using oxidative conditions, employing cerammonium nitrate (CAN) or 2,3–dichloro–5,6–dicyano–1,4–benzoquinone (DDQ). Benzyl ethers remain intact under these conditions. This was, for example, utilized in the synthesis of a 3–OH–unprotected lactosamine derivative, which is a disaccharide intermediate en route to the biologically relevant tetrasaccharide sialyl–Lewis[X] (cf. chapter 7).

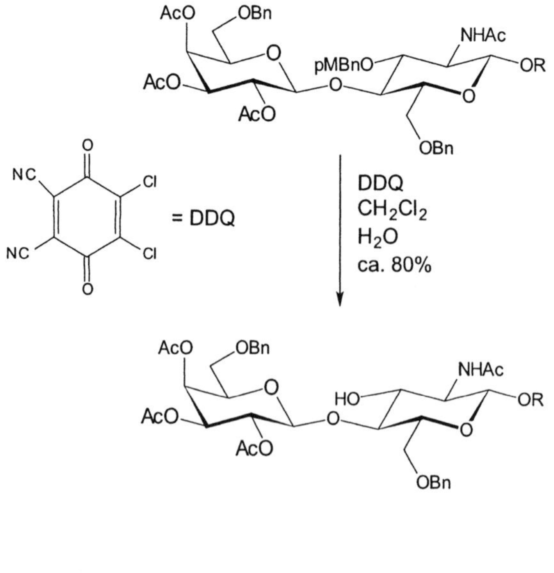

Allyl ethers

Allyl ethers are introduced under the same conditions as benzyl ethers, using allyl bromide instead of benzyl bromide. They are attractive protecting groups because they can be cleaved under very mild conditions which makes them fully compatible with almost all protection and deprotection as well as glycosylation procedures used in carbohydrate chemistry. To remove allyl ether protecting groups they are first isomerized employing a metal catalyst to form the respective 2–propenyl ether and then this labile enol ether is hydrolyzed under mild acidic conditions.

For allyl ether isomerization a wide variety of transition metal catalysts, including Pd (0), Rh (I), or Ir (II) can be used. Palladium on charcoal is often applied in refluxing methanol together with an organic acid. Alternatively, Wilkinson type catalysts such as [Ir(COD)(Ph$_2$MeP)$_2$]PF$_6$ and [(Ph$_3$P)$_4$Pd] can be used at room temperature. If strictly neutral conditions are required for the removal of allyl groups, for example when sensitive isopropylidene groups are present, a mixture of HgO and HgCl$_2$ in aqueous acetone can be used for enol ether cleavage instead of acid.

Selective etherifications

Selective etherification reactions with carbohydrates are far less feasible as selective acylations. As reducing monosaccharides are base–sensitive, direct etherification to obtain fully protected derivatives is not recommended and selectivity cannot be expected in this approach. Examples of monobenzylation of carbohydrates have been reported using phase transfer conditions. However, selective etherifications are most often achieved via the regioselective opening of intermediate dibutylstannylene acetals or tributylstannyl ethers, respectively. The use of dibutyltin oxide results in regioselective 2,6–di–O–benzylation of methyl glucoside.

Synthesis of methyl 2,6–di–O–benzyl–α–D–glucopyranoside.[12]

5 g (25.7 mmol) methyl α–D–glucopyranoside and 14.1 g (56.6 mmol) dibutyltin oxide are suspended in 300 ml of dry methanol in a round–bottomed flask. This is connected to a ro-

tary evaporator, in a 70°C water bath and the vacuum is adjusted so that small amounts of solvent are continuously removed by distillation. After 1 h the solvent is removed completely and the remaining stannane is dissolved in 40 ml benzyl bromide and heated to 90°C for 16 h. After that, the remaining benzyl bromide is removed under high vacuum by co–evaporation using water and toluene. The remaining crude product is purified by flash chromatography (dichloromethane–methanol, 20:1) to yield the desired product (82%) as a colorless syrup.

Synthesis of 3,6–di–*O*–alkylated mannosides can be achieved by conversion of the un-protected mannoside with bis(tributyltin)oxide ((Bu₃Sn)₂O) to form a tributylstannyl ether intermediate followed by reaction of the latter with the desired alkyl bromide. This reaction was used for preparation of octyl 3,6–di–*O*–allyl–α–D–mannoside, which forms the starting material for the synthesis of the branching mannotrioside occurring in the core region of *N*–glycans (cf. section 6.4). The octyl aglycone was used in this synthesis to substitute the two GlcNAc residues which link the branching mannotrioside to the peptide chain. Benzylation of the two remaining hydroxyl groups in the 3,6–di–allylated mannoside, followed by Ir–catalyzed removal of the allyl ethers furnishes a glycosyl acceptor which is mannosylated using acetyl protected mannosyl trichloracetimidate (cf. section 4.4) at the 3– and 6–positions. Reductive cleavage of the benzyl ether groups and deacetylation according to Zemplén gives rise to the desired mannotrioside. This could be used to probe the substrate specificity of a family of *N*–acetylglucosaminyltransferases which form part of the enzy-matic machinery which further elongates the oligosaccharide in glycoproteins.[13]

Trityl ethers

Trityl ethers can be used for the selective protection of primary hydroxyl groups because of the steric requirements of this protecting group. The trityl group is introduced under basic conditions, typically in pyridine using trityl chloride as the reagent. DBU or DMAP can be added to accelerate the reaction. Thus, methyl glucoside can be regioselectively protected at the 6–position in high yield by tritylation.

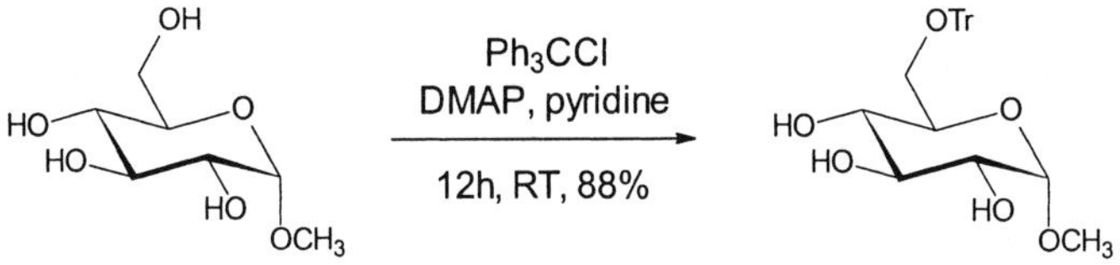

Cleavage of the trityl group is achieved under mild acidic conditions using mineral acids in ether. The conditions can often be mild enough to leave isopropylidene groups un-touched. The lability of trityl ethers may even be increased by using *p*–methoxy–substituted analogs, such as the monomethoxytrityl (Mmt) and dimethoxytrityl (Dmt) group, respec-tively. For each methoxy group introduced, acid–lability increases by about one order of magnitude. Mild Lewis acids or silica gel may now be enough to cleave the methoxy–substituted trityl group. The importance of trityl groups has diminished, however, by the introduction of the various silyl ethers, which are widely used in carbohydrate synthesis nowadays.

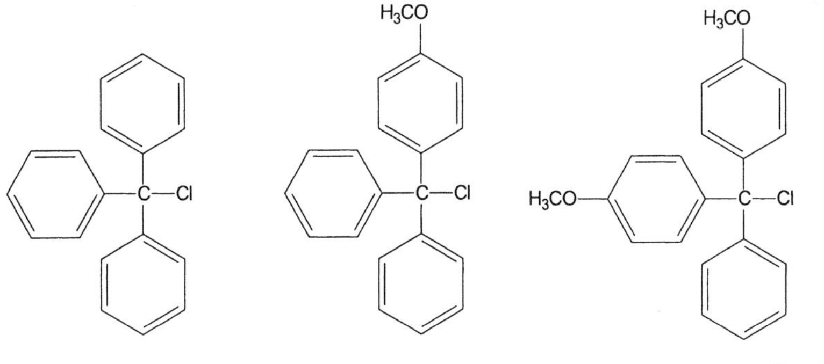

Trityl chloride   Monomethoxytrityl (Mmt)   Dimethoxytrityl (Dmt)
                              chloride   chloride

Silyl ether groups

Silyl ethers have gained considerable importance as protecting groups in natural product chemistry. They are used in a great number of variations, representing graded reactivity which is of significant importance for orthogonal protection. Silylations are easily per-formed using the respective silyl chlorides under basic catalysis, involving pyridine, imida-zole in DMF or hexamethyldisilazane ($Me_3SiNHSiMe_3$) in DMF. A very common silyl ether is the *t*–butyldimethylsilyl (TBDMS) group.

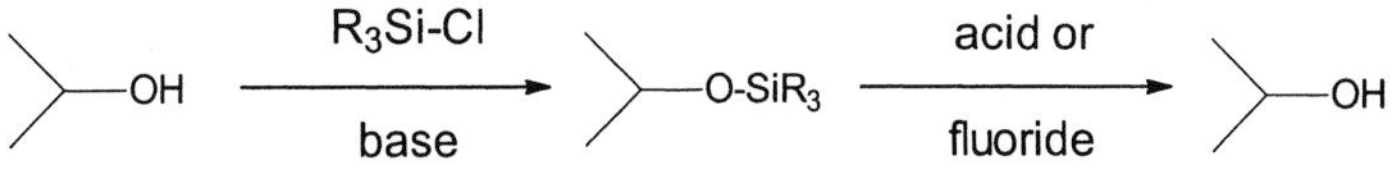

Applying TBDMS–chloride in pyridine allows selective protection of the primary hydroxyl group in pyranosides. In the case of methyl α–D–glucoside further silylation in DMF using imidazole as the catalyst affords the 2,6–di–O–silylated glucoside.

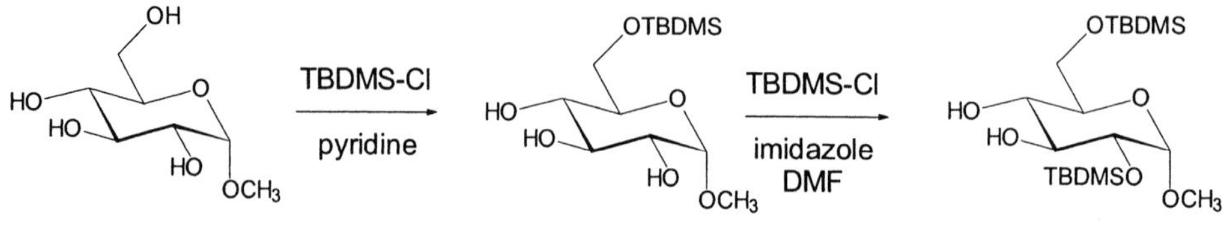

Regioselective silylation of the primary sugar hydroxyl group can be utilized to lock reducing pentoses into the furanose form. D–Arabinose, for example, in which the pyranose form predominates, can be converted to the 5–O–silylated furanose with *t*–butyldimethylsilyl chloride in pyridine. Acetylation of the remaining OH–functionalities can follow in the same reaction vessel to yield the fully protected furanose, which can be eventually converted to a 5–OH–free furanose by desilylation.

Synthesis of 1,2,3–tri–O–acetyl–5–O–*t*–butyldimethylsilyl–D–arabinofuranose.[14]

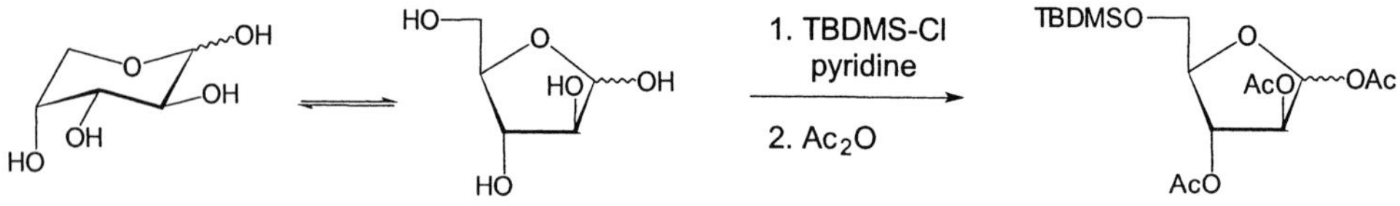

D-arabinopyranose   D-arabinofuranose

1,2,3-tri-O-acetyl-
5-O-*t*-butyldimethylsilyl-
D-arabinofuranose

10 g (66.6 mmol) arabinose are dissolved in 200 ml dry pyridine and treated with 11 g (73 mmol) TBDMS–Cl at 0°C. The reaction mixture is stirred at room temperature for 3.5 h, and then 30 ml of acetic anhydride are added and stirring is continued for 2 h. 200 ml saturated sodium hydrogen carbonate solution is added and the solution is extracted several times with dichloromethane. The combined organic phases are washed with water, dried over MgSO$_4$, filtered and the solvent is removed by evaporation. Flash chromatographic purification (ethyl acetate–petroleum ether, 1:1) gives 13.2 g (50%) of the desired furanose as almost colorless syrup.

Silyl ethers can be cleaved under acidic conditions. The most acid–sensitive silyl ether is the trimethylsilyl (TMS) ether, which is readily cleaved on silica gel. It is also the least sterically demanding and a fully TMS protected glycosides can, therefore, easily be obtained. By TMS protection monosaccharides become rather volatile, so that they can be investigated by gas–chromatography. Acid stability of silyl ethers increases in the following order:

TMS (trimethylsilyl) < TES (triethylsilyl) < TBDMS (*t*–butyldimethylsilyl) <
< TIPS (triisopropylsilyl) < TBDPS (*t*–butyldiphenylsilylchlorid)

Cleavage of silyl ethers can also be effected by fluoride ions, using tetrabutylammonium fluoride, for example. This is of great importance as it allows the removal of silyl ethers without cleavage of other acid–sensitive groups in the same carbohydrate derivative.

3.4   Acetal protecting groups

Cyclic acetals are used for simultaneous protection of two hydroxyl groups present in a sugar ring. They are normally formed in an acid–catalyzed reaction with ketones or aldehydes. The most common examples of acetal protecting groups are isopropylidene derivatives, arising from the reaction of a diol with acetone, and benzylidene derivatives, obtained by reaction of a diol with benzaldehyde. Other, less common cyclic acetals are cyclohexylidene, cyclopentylidene, DISPOKE and CDA acetals. All acetals can be removed under acidic conditions and are stable against bases and nucleophilic attack.

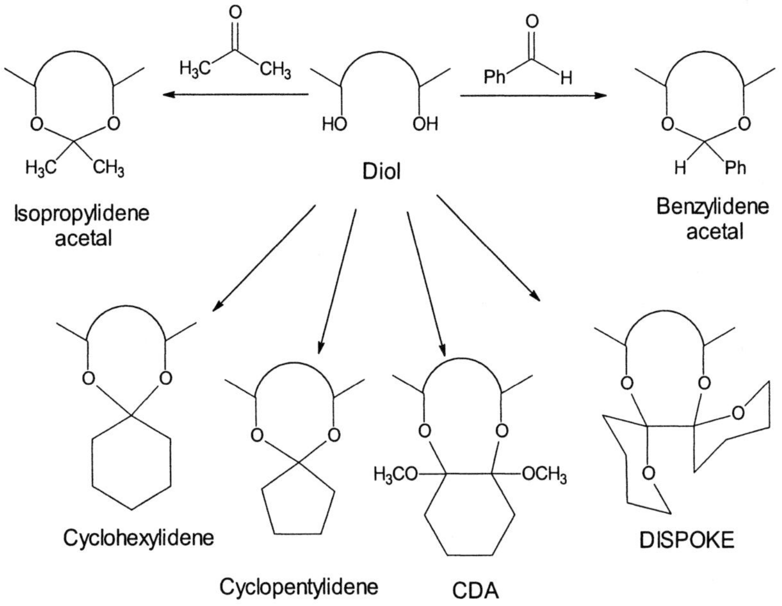

Isopropylidene
acetal

Diol

Benzylidene
acetal

Cyclohexylidene

Cyclopentylidene   CDA

DISPOKE

Isopropylidenation as well as benzylidenation can be used to protect certain arrangements of hydroxyl groups. For energetic reasons, acetone normally reacts with vicinal hydroxyl groups which are positioned in a *cis*–arrangement on the sugar ring, to give five–membered 1,3–dioxolanes, because the methyl axial substituent at the acetal center destabilizes the corresponding 1,3–dioxane. Benzylidene acetals, on the other hand, are most stable as six–membered rings with the phenyl substituent adopting the thermodynamically favored equatorial orientation. Five–membered benzylidene rings are only infrequently formed when the conformational preferences in dioxolanes are less pronounced.

Benzylidene acetals

Benzylidene acetals are mainly used in carbohydrate chemistry for 4,6–*O*–protection in hexopyranosides. In the case of methyl **gluco**side, the *trans*–bicyclic ring system is formed in the reaction with benzaldehyde, the analogous reaction with methyl **galacto**side yields the *cis*–fused 4,6–benzylidene derivative. *Trans*–annelated 4,6–*O*–benzylidene acetals of hexopyranosides represent a *trans*–decalinic system and are generally hydrolyzed faster than the corresponding *cis*–fused acetals which are analogous to a *cis*–decalinic system.

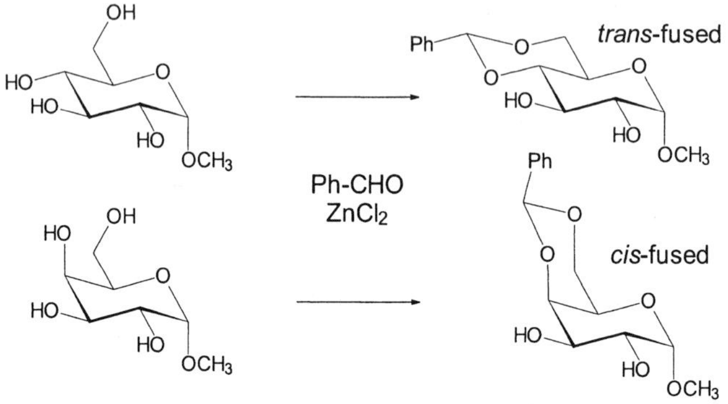

In the six–membered rings, such as the 4,6–*O*–benzylidene acetals, the phenyl substituent adopts the thermodynamically favored equatorial orientation. Therefore, after the benzylidenation of a sugar, in which a new stereocenter at the benzylidene group is created, only one diastereomeric product is observed. The classical recipe for the synthesis of 4,6–*O*–benzylidene protected carbohydrate derivatives starting with the free sugars uses benzaldehyde and zinc chloride as the catalyst.

Alternatively, benzaldehyde dimethylacetal is used as the reagent for benzylidenation together with a catalytic amount of *p*–toluensulfonic acid.

Synthesis of 4,6–*O*–benzylidene–D–glucopyranose.[15]

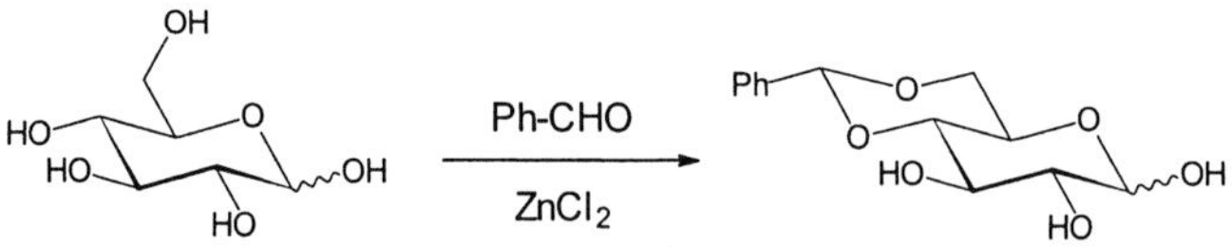

10 g of anhydrous glucose and 40 g of dry zinc chloride in 200 ml of pure benzaldehyde are stirred at room temperature for 4 h. Then, the reaction mixture is cooled to 0°C, 250 ml of water are added and it is stored at 4°C (refrigerator) for 1h. The resulting crystalline mass is filtered and washed with cold water to yield 7.5 g of the desired product.

Synthesis of methyl 4,6–O–benzylidene–α–D–glucopyranoside.[16]

Methyl α–D–glucopyranoside (9.7 g) and benzaldehyde dimethylacetal (7.6 g) are dissolved in 40 ml of dry DMF in a 250 ml round–bottomed flask. p–Toluenesulfonic acid (25 mg) is added and the flask is attached to a rotary evaporator, rotated, evacuated, and lowered into a water bath at approximately 60°C to remove the methanol which is formed during the reaction. After 1h, all the DMF is removed by evaporation, the temperature of the water bath being raised to approx. 100°C. Then the flask is removed from the evaporator, a solution of sodium hydrogen carbonate (1 g in 50 ml of water) is added to the residue, and the mixture is heated at 100°C until the product is finely dispersed. The mixture is cooled to room temperature, and the product filtered off, washed thoroughly with water, and dried for 4 h at 30°C and then overnight *in vacuo* over phosporus pentaoxide and paraffin wax to yield the desired benzylidene derivative in 82% yield. It can be recrystallized from isopropyl alcohol.

Regioselective 4,6–O–benzylidene protection at the non–reducing end of oligosaccharides is equally possible in one step from the free sugar employing the acid–catalyzed reaction with benzaldehyde dimethylacetal.

Synthesis of 4',6'–O–benzylidene–maltose.[17]

Maltose (40 g) is dissolved in 100 ml of dry DMF, benzaldehyde dimethylacetal (18 ml) and p–toluenesulfonic acid (50 mg) are added, and the mixture is rotated at 60–70°C on a rotary evaporator under a weak vacuum to remove the formed methanol. The reaction is continued for 6 h, with occasional replenishment of DMF. TLC analysis (ethyl acetate–methanol–water, 7:2:1) shows one major and several other minor products. The mixture is neutralized with basic ion–exchange resin, filtered, and coevaporated with toluene. The remaining syrup is subjected to water–ethyl acetate extraction, the organic phase being extracted twice with water, and the combined aqueous phases then washed three time with ethyl acetate. This procedure removes the more nonpolar byproducts from the aqueous phase, thus facilitating the purification procedure. The aqueous phase is concentrated by evaporation and purified by flash chromatography (ethyl acetate–methanol–water, 8:1:1) to yield the 4',6'–O–benzylidene derivative (20.4 g) in 43% yield.

p–Methoxybenzylidene, and *o*, *p*–dimethoxybenzylidene derivatives are sometimes used instead of their unsubstituted counterparts as they are more acid labile and can therefore be cleaved under milder conditions.

Selective opening of benzylidene acetals

Benzylidene acetals can be cleaved under acidic conditions or, alternatively, by hydrogenation. Moreover, benzylidene groups offer the advantage of several regioselective cleavage reactions opening a wide array of options for establishing orthogonal protecting group patterns. Three main reaction pathways are of importance for carbohydrate chemistry

(i)   radical oxidative opening using NBS to form the 4–OBz–6–OH derivative,

(ii)   reductive opening with lithium aluminum hydride and aluminum trichloride in ether to afford the 4–OBn–6–OH derivative, and

(iii)   reductive regioselective opening of the acetal employing sodium cyanoborohydride in HCl–ether leading to the 4–OH–6–OBn derivative.

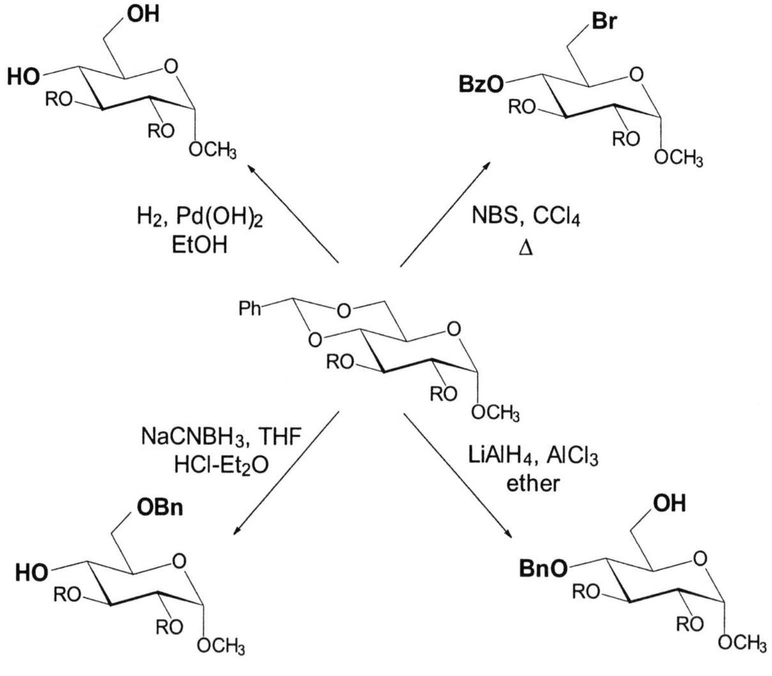

General procedure for the regioselective cleavage of 4,6–O–benzylidenated glycosides to form the respective 6–O–benzyl–4–OH derivatives. To a mixture of the acetylated, 4,6–O–benzylidene protected glycoside (e.g. 1.5 mmol) in dry THF (approx. 10 ml), NaCNBH$_3$ (13 mmol) is added. The mixture is stirred at room temperature and a saturated solution of HCl gas in dry ether is added in small portions until gas development ceases and the mixture remains acidic (approx. 30 minutes). Then it is concentrated by evaporation to about 2 ml, CH₂Cl₂ is added and the mixture is neutralized with satd. aq. NaHCO₃. The aqueous phase is extracted three times with CH₂Cl₂, the combined organic phases are washed with water, dried over MgSO₄, filtered and evaporated. The product can be obtained after chromatographic purification in high yields, often over 90%. This procedure can be equally well applied on di- or trisaccharides.

The mechanistic details of these reactions have not all been clarified as yet and the regiochemistry of the ring opening reactions with different derivatives are not always totally predictable. In methyl 2,3:4,6–di–*O*–benzylidene mannoside, the dioxolane ring can be chemoselectively cleaved without affecting the dioxane ring. However, cleavage with lithium aluminum hydride and aluminum trichloride gives the 3–*O*–benzylated mannoside in the case of the *endo*–benzylidene acetal and the opposite regioselectivity is observed in the same reaction starting with the *exo*–benzylidene acetal, which leads to the 2–*O*–benzylated product.[18]

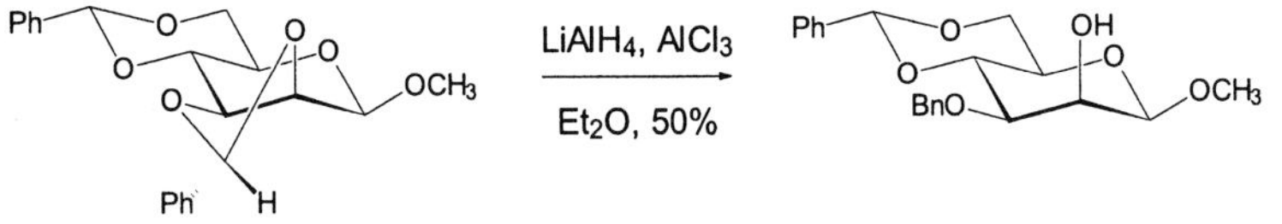

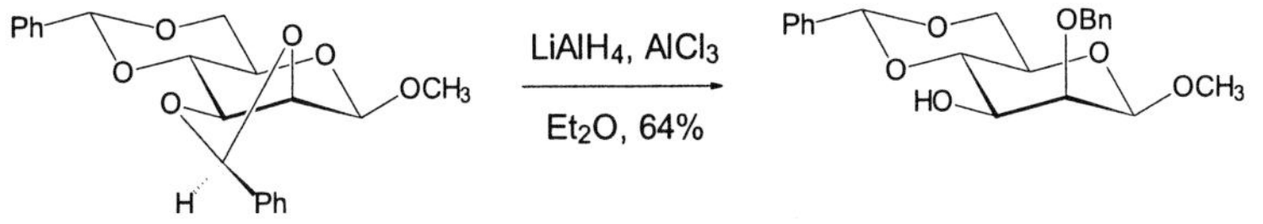

The regiochemistry of reductive cleavage of *p*–methoxybenzylidene acetals depends on the reaction conditions. By suitable choice of solvent and electrophile, the distribution of the two possible regioisomers can be controlled. Thus cleavage resulting in the *p*MBn ether of the more hindered hydroxyl group is accomplished with NaCNBH₃–TMSCl in acetonitrile, because trimethylsilylchloride is a sterically demanding electrophile, attacking at C–6. On the other hand, by using NaCNBH₃ with trifluoroacetic acid in DMF, cleavage occurs from the opposite direction to provide the *p*MBn ether of the less hindered hydroxyl.

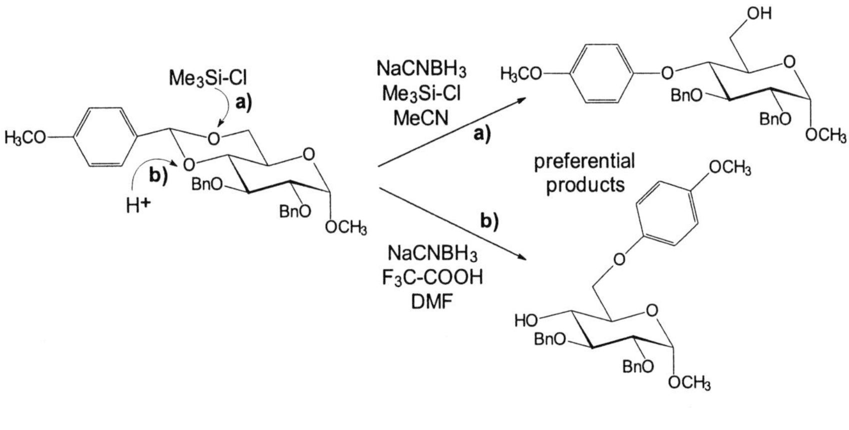

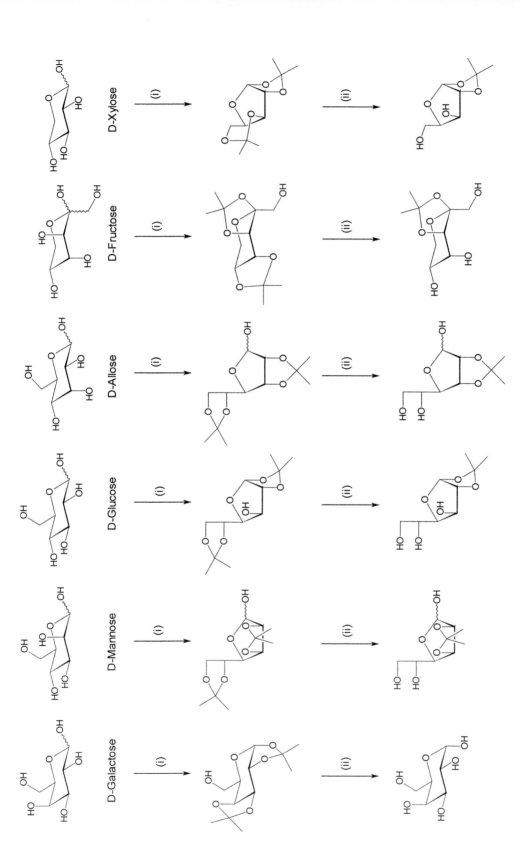

Several free monosaccharides can be converted by isopropylidenation ((i): H⁺, acetone) to useful building blocks which possess a single unprotected OH group, in a one step–reaction. The less stable isopropylidene group in each of the obtained diisopropylidene derivatives can be selectively cleaved to produce the respective sugars with 3 free hydroxyl groups ((ii): diluted HOAc).

Isopropylidene acetals

Isopropylidene acetals are often used for the protection of *cis*–positioned 1,2–diol groups in a saccharide. The isopropylidene group is introduced using dry acetone and an acid or Lewis acid catalyst, such as zinc chloride. Another reagent, which is often used, is 2,2–dimethyloxypropane, which undergoes a transacetalization reaction under catalysis of *p*–toluenesulfonic acid. Under these reaction conditions five–membered dioxolane rings are typically formed.

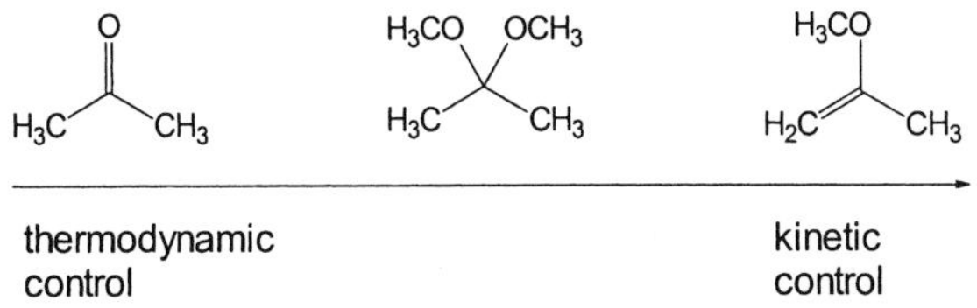

thermodynamic
control

kinetic
control

Less frequently 2–methyloxypropene is used to prepare acetonides allowing better kinetic control of the reaction than when acetone or its dimethylacetal is used. As a consequence, 1,3–dioxanes are formed in the acetalization reaction with 2–methyloxypropene, whereas the formation of dioxolanes is normally favored by the thermodynamic conditions. Based on this difference, glucose can be converted to a pyranose derivative fused with a dioxane or, alternatively, converted into a furanose derivative having two dioxolane rings.

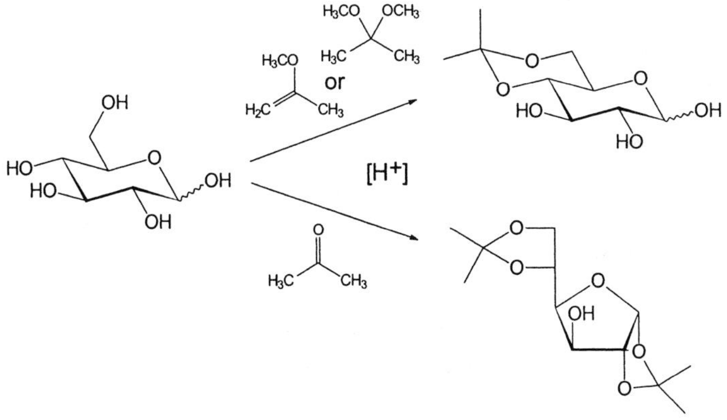

Moreover, using different (Lewis) acids to catalyze the reaction with acetone results in the formation of different isopropylidene building blocks, such as in the case of D–mannitol, for example. This can be converted to fully protected 1,2:3,4:5,6–tri–*O*–ispopropylidene–D–mannitol, using boric acid with sulfuric acid as the catalyst, whereas reaction in acetone with zinc chloride gives rise to a di–*O*–isopropylidene derivative in which the termini of the alditol chain are protected.

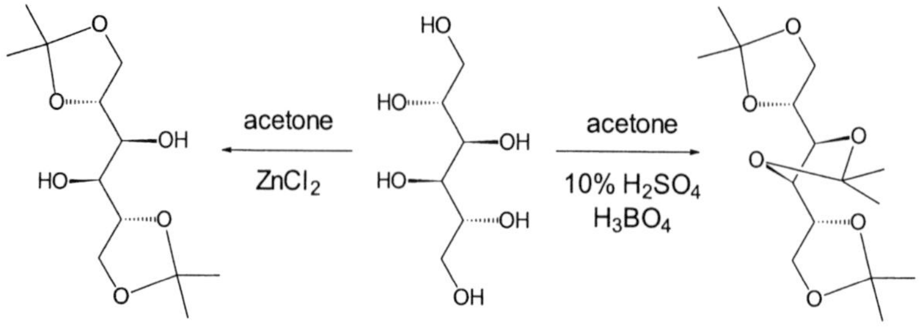

| 1,2:5,6-di-O-isopropylidene-D-glucitol | D-mannitol | 1,2:3,4:5,6-tri-O-isopropylidene-D-mannitol |

Acetalization of unprotected pyranoses, furanoses, and alditols leads to useful isopropylidene protected building blocks with one unprotected hydroxyl group in many cases. Thus, galactose reacts with acetone to give the 6–OH–free pyranose (diisopropylidene–galactose), which is often used as reactive glycosyl acceptor to test a particular glycosylation procedure. Mannose and glucose, on the other hand, form di–O–isopropylidene furanose derivatives, because in the furanose form the formation of the less stable *trans*–fused dioxolanes is avoided. Also allose, fructose, and xylose form di–O–isopropylidene derivatives in one step from the free sugars.

Selective hydrolysis of isopropylidene derivatives

Isopropylidene acetals are cleaved under acidic conditions, using, for example, different dilutions of acetic acid, sulfuric acid in aqueous methanol, silica gel or acidic ion exchange resin. The exact cleavage conditions have to be worked out for every isopropylidene derivative, especially when a discrimination from other acid–sensitive groups present in the molecule is required. Moreover, selective cleavage of only one or two of several ispropylidene groups in a sugar can be accomplished as not all isopropylidene moieties in a carbohydrate derivative are of equal stability. This can be demonstrated by 1,2:3,4:5,6–tri–O–isopropylidene–D–glucitol, obtained in one step from D–glucitol. The two end isopropylidene groups in this tri–O–isopropylidene derivative can be selectively removed with aqueous acid leading to 3,4–O–isopropylidene–glucitol.

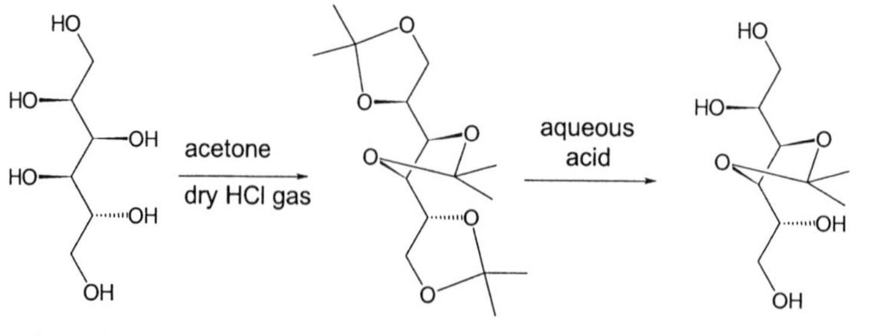

| D-glucitol | 1,2:3,4:5,6-tri-O-isopropylidene-D-glucitol | 3,4-O-isopropylidene-D-glucitol |

In pyranoses and furanoses a 1,2–*O*–isopropylidene group is generally more resistant to acidic hydrolysis than the same group at any other position. Consequently, most 1,2:5,6–di–*O*–isopropylidene acetals of aldohexoses can be partially hydrolyzed to the 1,2–*O*–isopropylidene derivatives. For sugars in which the acetal function does not involve the anomeric center, a 1,3–dioxolane *cis*–fused to a furanose or a pyranose is more stable than a 1,3–dioxolane which involves a side chain. Moreover, the five–membered dioxolanes are clearly more stable than the six–membered dioxanes.

Silyl acetals

One silyl acetal in particular has gained importance as a protecting group in carbohydrate chemistry, the tetra–isopropyl–disiloxanylidene (TIPDS) group, which protects *cis*– as well as *trans*–1,2–diols. The TIPDS group is introduced with the difunctional reagent, 1,3–dichloro–1,1,3,3–tetra–isopropyl–disiloxane using weakly basic catalysis, such as in pyridine or with imidazole in DMF. Normally, under these reaction conditions the eight–membered ring is formed under kinetic control. This can, in turn, rearrange to the thermodynamically more stable seven–membered TIPDS acetal under conditions of acidic catalysis such as those prevailing when *p*–toluenesulfonic acid is used.

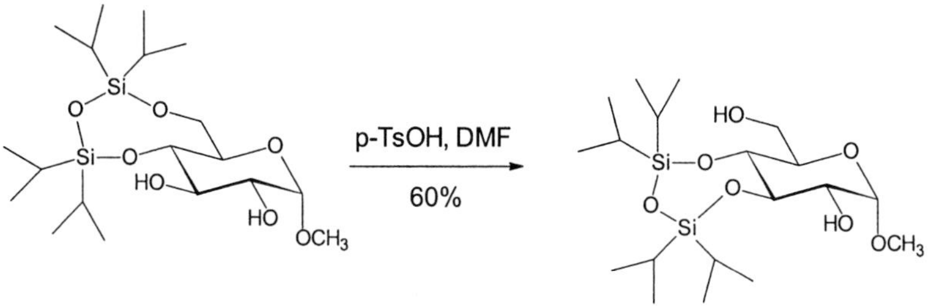

TIPDS is quite frequently employed in Kdo chemistry for the simultaneous and regioselective protection of the 7– and 8–OH groups in the terminal chain of this sugar.

Synthesis of methyl (allyl–*O*–3–deoxy–7,8–tetraisopropyldisiloxane–1,3–diyl–α–D–manno–2–octulopyranoside)onate. [19]

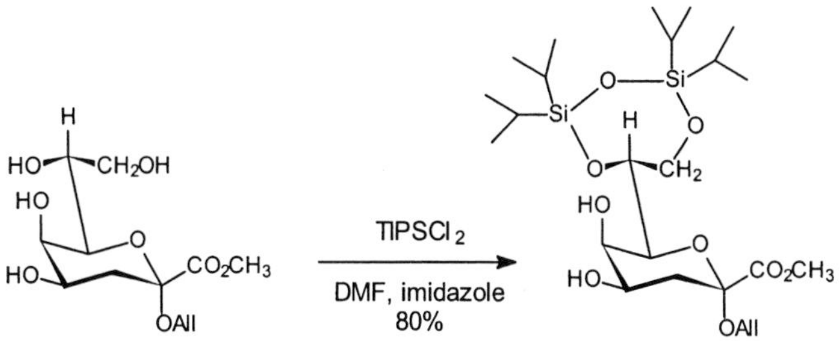

Methyl (allyl–*O*–3–deoxy–α–D–manno–2–octulopyranoside)onate (5.4 mmol) is dissolved in 25 ml of dry DMF, cooled to –10°C and then a solution of 5.4 mmol TIPDSCl$_2$ and 13 mmol imidazole in 20 ml of dry DMF is added dropwise. After 4h more TIPDSCl$_2$ solution (1.6 mmol TIPDSCl$_2$ and 3.9 mmol imidazole in 7 ml DMF) is added and the temperature of the reaction mixture is allowed to rise to 0°C. The reaction is completed after another 5 h of stirring at 0°C. Excess reagent is destroyed by adding 7 ml of methanol and the mixture is then evaporated and co–evaporated several times with toluene under high vacuum. The residue is dissolved in dichloromethane, washed with satd. aq. NaHCO$_3$ solution and then with water. The organic layer is separated and dried (MgSO$_4$), filtered and evaporated *in vacuo*. After chromatographic purification the syrupy product is obtained in 80% yield.

TIPDS acetals survive hydrogenation and mild glycosylation conditions. They are cleaved, as all silyl ethers, under mild acidic conditions and also by fluoride. This was favorably employed in glycosylation reactions using acetylated glycosyl fluorides. Activation of the glycosyl donor with the Lewis acid boron trifluoride etherate leads to the release of fluoride, which attacks the 4,6–*O*–TIPDS protecting group of the benzyl mannoside, which is shown. This leads to regioselective opening of the silyl acetal and allows the formation of the respective 6–linked disaccharide.[20]

DISPOKE and CDA

For steric reasons benzylidene and isopropylidene acetals are less suited for the selective protection of *trans*–diols in carbohydrates. This can be better accomplished with two relatively new protecting groups, the so–called dispiroketal (DISPOKE) and cyclohexane–1,2–diacetal (CDA) groups. The reagents used in the acid–catalyzed protection reaction reagents are 3,3',4,4'–tetrahydro–6,6'–bis–*2H*–pyran and 1,1,2,2–tetramethoxycyclohexane. Both acetals are removed by transacetalization with ethylene glycol.

DISPOKE and CDA protecting groups are compatible with a wide range of other protecting groups such as Bn, *p*MBn, TBDMS, and acyl groups. They are a good complement to benzylidene and isopropylidene acetals, as DISPOKE protects *trans*–1,2–diols in preference to *cis*–1,2–diol arrangements and 1,2–*cis*–diols in preference to 1,3–diols. Besides DISPOKE, only the TIPS group protects *trans*–diols rather than *cis*–diols.

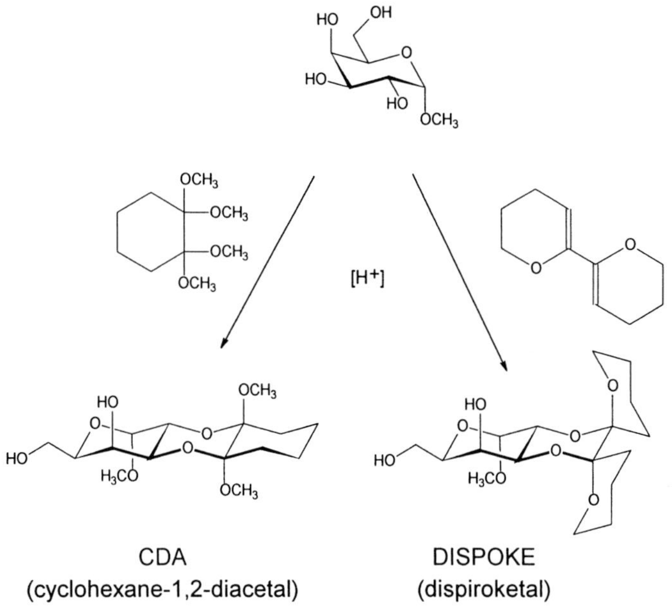

CDA
(cyclohexane-1,2-diacetal)

DISPOKE
(dispiroketal)

Acyclic acetals

A number of acyclic *O,O*–acetals are known as protecting groups, of which the tetrahydropyranyl (THP) group is the most frequently used in carbohydrate chemistry. Acid catalyzed addition of primary as well as secondary alcohols to dihydropyran in dichloromethane is used for the preparation of tetrahydropyranyl acetals. The reaction proceeds by protonation of the β–carbon atom of the enol ether oxygen to generate a highly electrophilic oxocarbenium ion as shown. This is then attacked by the alcohol at the carbon atom adjacent to the oxygen. In carbohydrate chemistry THP acetals have been used to selectively protect the primary hydroxyl group of a sugar.

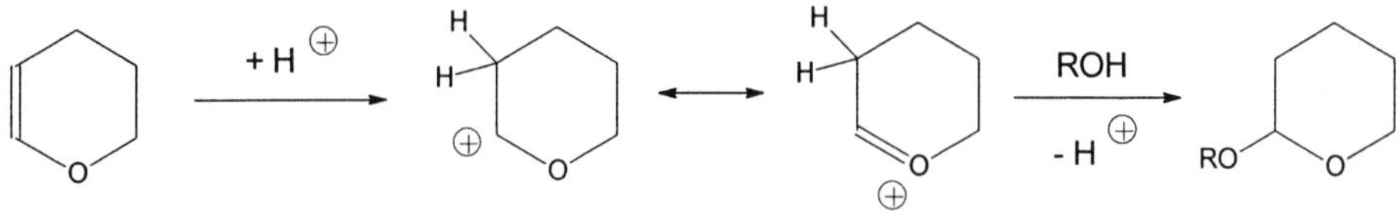

However, two diastereomers are formed upon THP protection because a new stereogenic center is formed at the tetrahydropyranyl residue. This causes increased complexity of the NMR spectra of the products and, therefore, THP acetals never became very popular protecting groups in carbohydrate chemistry.

Other acyclic *O,O*–acetals used as protecting groups are

MOM   Methoxymethyl–,
MEM   Methoxyethoxymethyl–
MTM   Methylthiomethyl– (a *O, S*–acetal), and
BOM   benzyloxymethyl.

All these protecting groups are introduced using the respective chlorides as reagents and a base, such as NaH or diisopropylethyl amine to form the alkoxide ion from the sugar alcohol. They are labile under mildly acidic conditions. THP ethers can be cleaved by HOAc–THF–H_2O (4:2:1) at 45°C, conditions which do not cleave MEM, MTM, or MOM ethers.

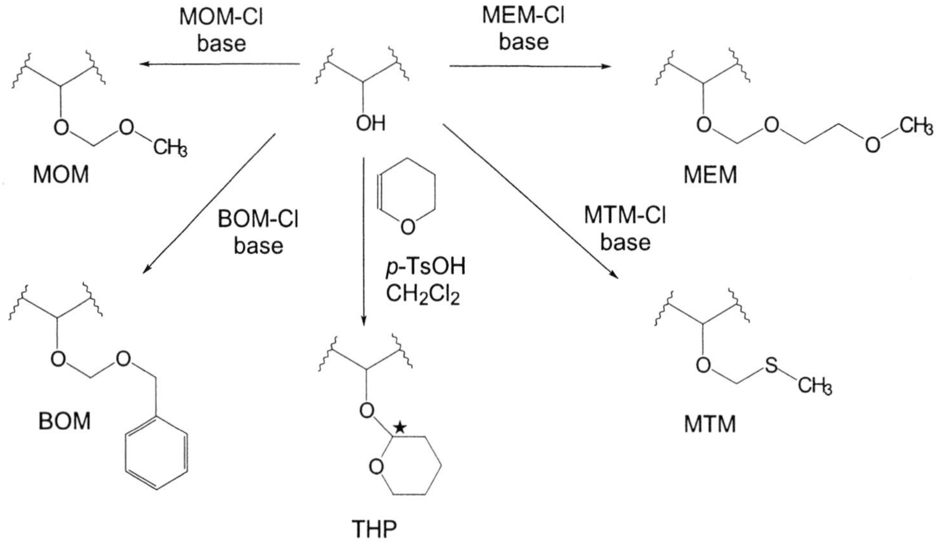

Acyclic acetal protecting groups used for the protection of alcohols.

3.5   Orthoesters as protecting groups

Orthoester formation in carbohydrates involves two adjacent, *cis*–positioned hydroxyl groups of the sugar ring. Glycosyl orthoesters are useful intermediates in the synthesis of sugar building blocks, offering a variety of options for their formation as well as for their cleavage. 1,2–Glycosyl orthoesters are often observed as byproducts in glycosylation reactions, especially in the mannose series. They can, for example, arise from the nucleophilic attack of an alcohol at the dioxolane ring carbon atom of the oxocarbenium ion, which is formed as an intermediate in 1,2–*trans*–glycosidations (cf. section 4.1). To synthesize 1,2–glycosyl orthoesters, the corresponding glycosyl bromide is treated with 1,3–dimethyl–pyridine in dichloromethane.

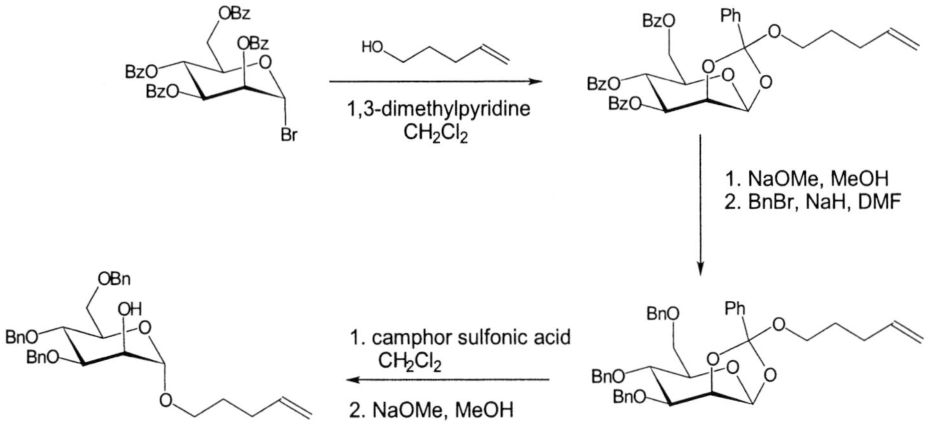

2–*O*–Acetyl as well as 2–*O*–benzoyl protected glycosyl bromides are used as starting material in this procedure. The resulting orthoacetates and orthobenzoates, respectively, survive a variety of functional group conversions at the 3–, 4–, and 6–postion of the sugar ring. Isomerization of the orthoesters which have been protected, then gives rise to glycosides, in which the 2–OH group is orthogonally protected in relation to the others. *n*–Pentenyl glycosides, which can serve as glycosyl donors (cf. section 4.6), have therefore often been prepared via orthoester intermediates.[21]

Orthoesters can also be cleaved with acetic acid to form the corresponding 2–*O*–acyl protected reducing sugars. These, in turn, can be converted to glycosyl trichloroacetimidates (cf. section 3.1 and 4.3) with complex functionalization at the sugar ring.

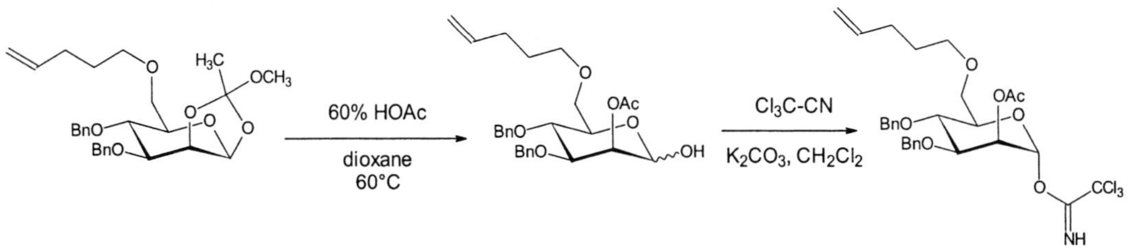

Other sugar orthoesters are prepared from unprotected glycosides employing diverse orthoesters of simple alcohols under acid catalysis. Triethyl and trimethyl orthoformate, orthoacetate, and orthobenzoate, respectively, are reagents frequently employed. The reason for the common use of sugar orthoesters is that the dioxolane ring may be cleaved regioselectively under acidic conditions, to yield a product in which the axial OH is acyl protected and the equatorial hydroxyl left free. Starting from methyl fucoside, a selectively 3–OH–unprotected building block can thus be prepared in a one–pot reaction sequence.

Synthesis of methyl 2,4–di–O–benzoyl–6–deoxy–α–L–galactopyranoside.[22]

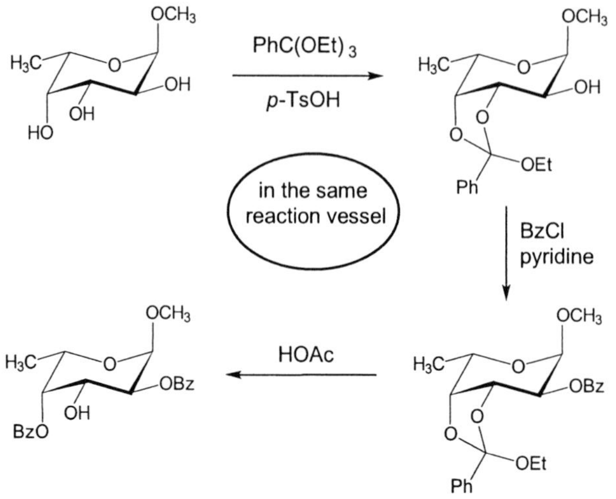

To a solution of 2 g methyl α–L–fucoside (methyl 6–deoxy–α–L–galactopyranoside) in 40 ml dry dichloromethane, 7.5 ml triethyl orthobenzoate and a catalytic amount of p–toluenesulfonic acid are added and the reaction mixture is stirred at room temperature for approx. 30 minutes until the sugar orthoester has formed (observed using TLC). Then, the reaction mixture is added dropwise to a solution of 4 ml benzoyl chloride in 20 ml dry pyridine at 0°C. Stirring at room temperature is continued until the benzoylation reaction is complete (observed using TLC). 15 ml of acetic acid are added and after complete cleavage of the orthoester, the mixture is poured into satd. aq. sodium hydrogen carbonate solution and extracted several times with dichloromethane. The combined organic phases are washed with water, dried (MgSO₄), filtered and evaporated. The resulting solid can be crystallized from toluene to yield 3 g (70%) of the desired product in the form of colorless needles.

3.6 *N–*Protection for amino sugars

The amino groups of 2–amino–2–deoxy amino sugars are mostly *N*–acetylated in the majority of naturally occurring oligosaccharides, as in the case of the important monosaccharide constituents *N*–acetylglucosamine and *N*–acetylgalactosamine. On the other hand, also *N*–unprotected amino sugars are components of biologically active glycoconjugates, for example in GPI–anchors were *N*–unprotected amino sugars coexist with their *N*–acetylated analogs.

N–Acetyl protected amino sugars usually do not undergo reactions at the nitrogen atom. They are, however, only cleavable under extreme basic conditions, and therefore *N*–acetylation normally is not used as temporary protection. In cases, where a *N*–acetylated product is desired *N*–acetylation appears as suitable *N*–protection, however, *N*–acetylated amino sugars are particularly unsuited as glycosyl donors, because they tend to form stable oxazoline derivatives during the glycosylation step (cf. section 4.7).

N–chloroacetylated amino sugars are an alternative to the N–acetylated analogs as N–chloroacetylated monosaccharide tetraacetates can be directly used as glycosyl donors in a $FeCl_3$–catalyzed glycosidation reaction. They are more reactive glycosyl donors than acetamides or oxazolines. After the glycosylation step the N–chloroacetyl group can be converted to the N–acetyl group by reductive dehalogenation using Zn in acetic acid–THF.

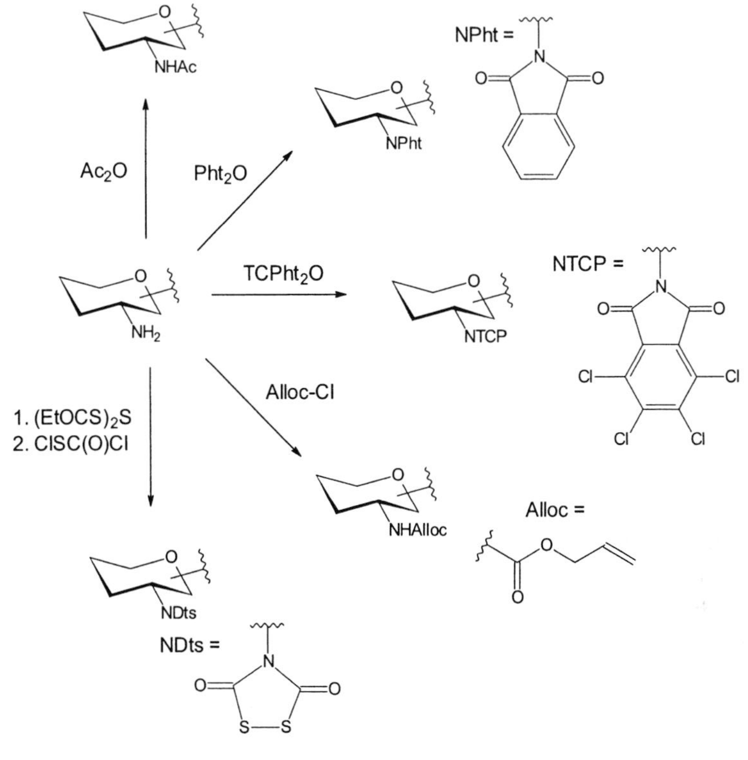

In order to avoid the formation of 1,2–oxazoline sugars during glycosylation, 2–phthalimido sugars are a widely used as N–protecting groups in sugar chemistry. The phthalimido group is introduced by treatment of the free glycosylamine with phthalic anhydride. However, removal of the phthalimido group is often connected with problems and therefore, the phthalimido group is more and more replaced by more electron–withdrawing derivatives in order to facilitate the cleavage step.[23] The N–tetrachlorophthalimido (TCP) group was shown to be particularly successful in this regard. It can be cleaved under very mild conditions such as with ethylenediamine or sodium borohydride.

The dithiasuccinoyl (Dts) group is another modern group for N–protection of glycosyl-amines. It combines the advantages of easy installation, stereoselective formation of β–glycosides, effectiveness in solid phase synthesis and extremely mild conditions for re-moval. Dts is introduced by reaction of the glycosylamine hydrochloride with either S–

carbomethyl O–ethyl dithiocarbonate in methanol or bis(ethoxythiocarbonyl)sulfide in aqueous ethanol, followed by subsequent acetylation with acetic anhydride in pyridine. The obtained ethoxythiocarbonyl intermediate is then treated with (chlorocarbonyl)sulfenyl chloride, and in the accompanying cyclization reaction the dithiasuccinoyl group is formed with the extrusion of ethyl chloride and hydrogen chloride.

The Dts protecting group can be cleaved under thiolytic conditions, using thiols such as dithiothreitol. This reaction is catalyzed by tertiary amines such as N,N–diisopropylethylamine, which also reduces azido groups. Dts can, however, also be removed with sodium borohydride without affecting azido groups.

The use of allyloxycarbonyl amides as amino protection allows deprotection under almost neutral conditions. Isomerization of the double bond with for example tetrakis(triphenylphosphine) palladium and a mild base or the use of an aqueous or two phase system employing Pd(0) and diethylamine removes the allyloxycarbonyl (Alloc) protecting group. Alloc is introduced with allyloxycarbonyl chloride or allyl 1–benzotriazoyl carbonate and triethylamine in THF. Tetraacetylated, N–Alloc protected amino sugar derivatives can be used as glycosyl donors.

Azides can be considered as masked amino groups and can thus serve as N–protection in a sense. As non–participating groups they also allow the synthesis of 2–amino–2–deoxy–α–D–glycosides. 2–Deoxy–2–azido derivatives are obtained by multi–step syntheses, for example by nucleophilic substitution reaction of suited trifluoromethanesulfonate derivatives as shown or by azidonitration of glycals (cf. section 5.4). Azido groups are easily reduced to amino groups by catalytic hydrogenation. However, when C=C–double bonds or benzyl groups are present in the same molecule which would also be affected under these conditions, hydrogen sulfide in pyridine allows the chemoselective reduction of azide in the presence of these other reducable groups.

Conversion into glycosyl donors possible

3.7 Other protecting groups

In protecting carbohydrates, not only do protecting groups have to be selected for every individual case, but, more importantly, a reasonable protecting group strategy has to be planned prior to the synthesis. Experience shows that it is easier to use as few different protecting groups as possible, than to have a large number of orthogonal protecting groups established in one derivative. Also is it advantageous to have a uniform protecting group pattern before the last deprotection step of a multi–step synthesis. In addition, one should consider that a particular reaction might also proceed favorably without protecting groups, making use of regioselectivities for example. In this respect, no protecting group can be considered to be the best one.

In spite of these considerations it might still be useful to evaluate the use of other unusual, less commonly used protecting groups to solve a particular protection problem. Changing the protection pattern in a somewhat unusual way might also improve the reactivity of the system as well as its orthogonal potential. Two often overlooked possibilities are mentioned in the following sections, these are photo–cleavable protecting groups and the use of enzymes for selective protection and deprotection, respectively.

Photosensitive protecting groups

Photosensitive protecting groups contain a chromophore, which shows high chemical stability but is activated by light of a certain wavelength. Most often used in carbohydrate chemistry is the *o*–nitrobenzyl group, which is employed in the form of its ether, ester, acetal, carbamate and even carbonate.

The removal of 2–nitrobenzyl groups is based on a photoinduced, intramolecular oxidation–reduction reaction.

Interestingly, irradiation of 2–nitrobenzylidene acetals under oxidative conditions can result in regioselective cleavage of the dioxolane ring and synthesis of a selectively 2–nitrobenzoyl protected derivative in which the *o*–NBz group may eventually be removed under photochemical conditions.

Use of enzymes for selective protection and deprotection

Ester–cleaving hydrolases from different sources have been successfully used in carbohydrate chemistry both for regioselective esterification and for selective hydrolysis of sugars esters. Using enzymes offers the advantage of high selectivity and mild reaction conditions. Moreover, hydrolases are less sensitive enzymes, often working even in organic solvents in which protected sugars are soluble. They can be purchased and many of them are inexpensive. Regioselectivity differs from enzyme to enzyme and has to be experimentally established for every reaction.

The enzymes used have generally been lipases such as porcine pancreatic lipase (PPL) or lipases obtained from microorganisms, such as from *Candida cylindracea* (lipase–CC) or from *Pseudomonas fluorescens* (lipase–P). Also subtilisin, an extracellular serine protease from *Bacillus subtilis* could been employed for regioselective esterifications in a large number of examples.

In enzyme–catalyzed hydrolysis of peracylated sugars, esters of primary OH groups are preferentially hydrolyzed, as shown with peracylated methyl glucoside derivatives.

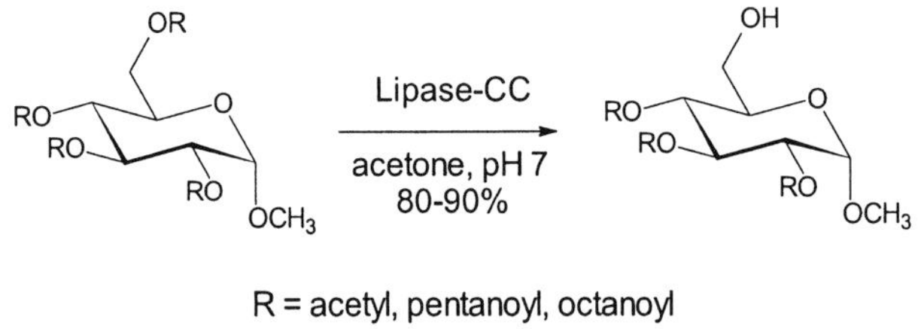

R = acetyl, pentanoyl, octanoyl

Surprisingly, tri–*O*–acetyl–D–glucal could be regioselectively hydrolyzed in slightly buffered ethyl acetate solution to give 4,6–di–*O*–acetyl–D–glucal in 90% yield using lipase–P.

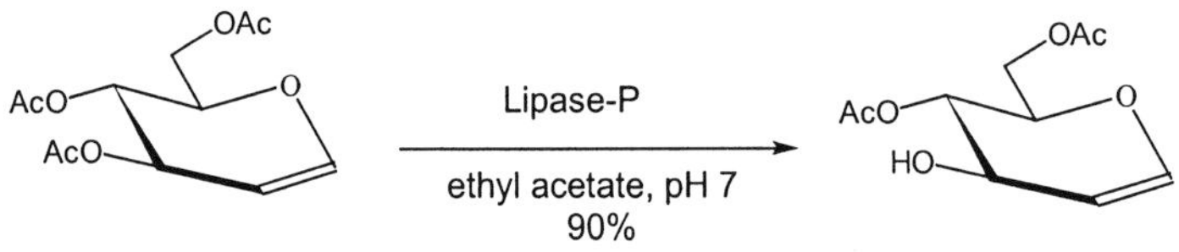

Acylation reactions are most often carried out as transesterification reactions using the respective esters for the introduction of a wide variety of alkanoyl residues. Thus D–glucal can be catalytically 6–*O*–acetylated with vinyl acetate using lipase–CC.

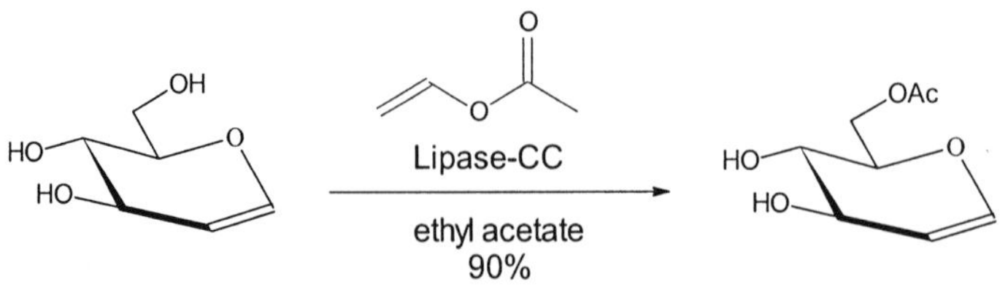

3.8 Table of common carbohydrate protecting groups

Protecting group	Method of attachment	Method of removal	Stability against base	Stability against acid	Stability against H₂, [cat]	Stability against hydride	Stability against oxidation
Acetyl (Ac)	Ac$_2$O or AcCl in pyridine	NaOMe, MeOH	–	+	+	–	+
Benzoyl (Bz)	BzCl in pyridine	NaOMe, MeOH	–	+	+	–	+
Pivaloyl (Piv)	PivCl in pyridine	KOH in MeOH	–	–	+	–	+
Chloroacetyl (AcCl)	ClAc–Cl in pyridine	Thiourea in EtOH/pyridine	–	+	+	–	+
Benzyl (Bn)	BnBr, NaH, DMF	Pd–C, H$_2$	+	+	–	+	–
p–Methoxy-benzyl (pMBn)	p–MeOBnCl, NaH, DMF	DDQ in CH$_2$Cl$_2$ or CAN	+	–	–	+	–
Allyl (All)	AllBr, NaH, DMF	[cat], e.g. (Ph$_3$P)$_3$RhCl	+	+	–	+	+/–
Trityl (Tr)	Ph$_3$C–Cl in pyridine	AcOH–H$_2$O	+	–	–	+	+
TBDMS	TBDMS–Cl, imidazole/DMF or in pyridine	AcOH–H$_2$O or TBAF	+	–	+	+	+
Isopropylidene	acetone, H$_2$SO$_4$, CuSO$_4$, or Me$_2$C(OMe)$_2$, p–TsOH	AcOH–H$_2$O or TFA/THF/H$_2$O	+	–	+	+	+
Benzylidene	PhCHO, ZnCl$_2$ or PhCH(OMe)$_2$, p–TsOH	AcOH–H$_2$O or Pd–C, H$_2$ or TFA/THF/H$_2$O	+	–	–	+	–
THP	DHP, p–TsOH, CH$_2$Cl$_2$	TFA/THF/H$_2$O	+	–	+	+	+

References

1.   U. Zehavi, N. Sharon (1972) *J. Org. Chem.* 37, 2141.
2.   R. R. Schmidt, M. Stumpp (1983) *Liebigs Ann. Chem.* 1249.
3.   R. Boss, R. Scheffold (1976) *Angew. Chem. Int. Ed. Engl.* 15, 558.
4.   K. Jansson, S. Ahlfors, T. Frejd, J. Kihlberg, G. Magnusson, J. Dahmén, G. Noorl, K. Stenvall (1988) *J. Org. Chem.* 53, 5629.
5.   M. L. Wolfrom, A. Thompson (1963) *Methods Carbohydr. Chem.* 2, 211.
6.   G. Zemplén, E. Pascu (1929) *Ber. Dtsch. Chem. Ges.* 62, 1613.
7.   Th. K. Lindhorst, J. Thiem (1991) *Carbohydr. Res.* 209, 119.
8.   J. Zhang, P. Kováč (1999) *J. Carbohydr. Chem.* 18, 461.
9.   A. V. Nikolaev, I. A. Ivanova, V. N. Shibaev, N. K. Kochetkov (1990) *Carbohydr. Res.* 204, 65.
10.   C. Kieburg, K. Sadalapure, Th. K. Lindhorst (2000) *Eur. J. Org. Chem.* 2035.
11.   H. Paulsen, H. Behre, C.-P. Herold (1970) *Topics Curr. Chem.* 14, 472.
12.   M. A. Nashed, L. Anderson (1976) *Tetrahedron Lett.* 3503.
13.   T. Ogawa, K. Katano, K. Sasajima, M. Matsui (1981) *Tetrahedron* 37, 2779.
14.   Th. K. Lindhorst, J. Thiem, unpublished.
15.   N. K. Richtmeyer (1962) *Methods Carbohydr. Chem.* 1, 107.
16.   M. E. Evans (1972) *Carbohydr. Res.* 21, 473.
17.   Th. K. Lindhorst, C. Braun, S. G. Withers (1995) *Carbohydr. Res.* 268, 93.
18.   Z. Szurmai, L. Balatoni, A. Lipták (1994) *Carbohydr. Res.* 254, 301.
19.   P. Kosma, G. Schulz, H. Brade (1989) *Carbohydr. Res.* 190, 191.
20.   T. Ziegler, E. Eckhardt, G. Pantkowski (1994) *J. Carbohydr. Chem.* 13, 81.
21.   U. E. Udodong, R. Madsen, C. Roberts, B. Fraser-Reid (1993) *J. Am. Chem. Soc.* 115, 7886.
22.   Th. K. Lindhorst, J. Thiem (1990) *Liebigs Ann. Chem.* 1237.
23.   J. Debenham, R. Rodebaugh, B. Fraser-Reid (1997) *Liebigs Ann. Chem.* 791.

4   *O*–Glycoside synthesis

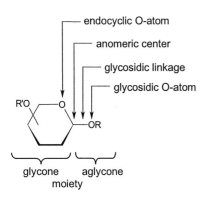

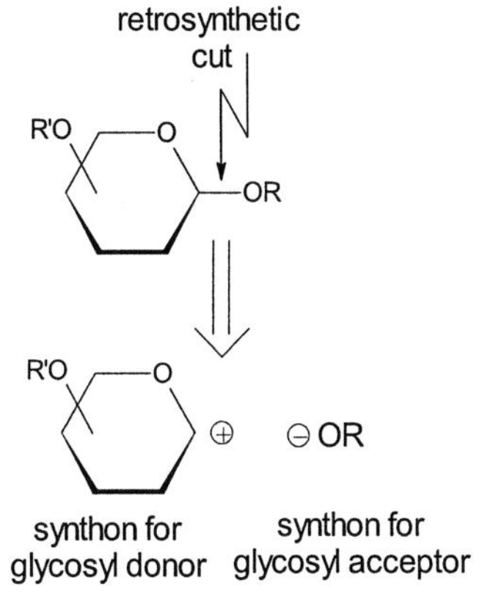

endocyclic O-atom
anomeric center
glycosidic linkage
glycosidic O-atom

R'O—O

—OR

glycone   aglycone
moiety

retrosynthetic
cut

R'O—O

—OR

R'O—O

⊕   ⊖OR

synthon for   synthon for
glycosyl donor   glycosyl acceptor

The synthesis of either naturally–occuring or ana-logs of oligosaccharide structures in which certain naturally established groups have been changed or deleted, is of great interest, for example for the in-vestigation of the three–dimensional structure of defined oligosaccharides, for testing their biological effects and for mapping saccharide binding sites in carbohydrate–binding proteins such as enzymes and lectins. Consequently, oligosaccharide chains of glycoconjugates are important targets and leading structures.

Generally, a glycoside is a cyclic acetal of the structure shown, with a glycone moiety and an agly-cone moiety bound to it at the anomeric center. The term glycoside, when no other characteristics are specified, normally refers to an *O*–glycoside, as in naturally occurring examples. Carbohydrate chem-istry, however, also deals with the synthesis of *N*–, *S*–, and *C*–glycosides, which are not discussed in detail in this book. A retrosynthetic analysis cleaves a glycoside at the glycosidic linkage into an electro-philic glycosyl donor synthon and a nucleophilic glycosyl acceptor equivalent. During a glycosylation reaction glycosidation of the hemiacetal into an acetal (glycoside) occurs, while the alcohol compo-nent is glycosylated during the same conversion.

Thus, oligosaccharide synthesis requires a glyco-sylation methodology which implies (i) the activation of a sugar into a suitable glycosyl do-nor equipped with a leaving group 'L' at the anomeric center, and (ii) the efficient and stereoselective coupling to the glycosyl acceptor, which is promoted by a suitable activator to lead to an oligosaccharide or a disaccharide in the simplest case. The glycosyl acceptor must be sufficiently protected prior to the glycosylation step.

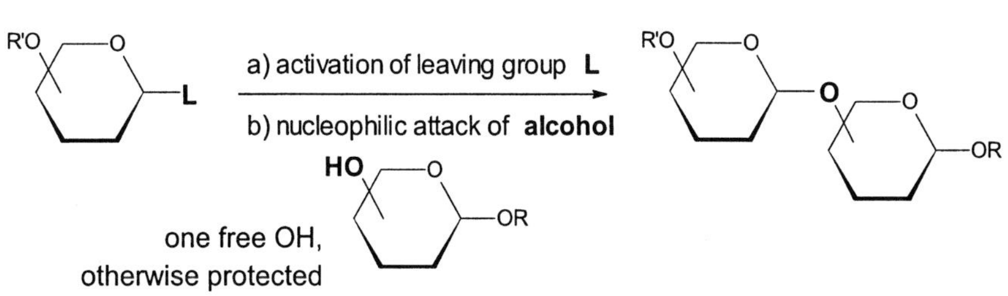

R'O—O

—L

a) activation of leaving group **L**
b) nucleophilic attack of **alcohol**

HO—O

—OR

one free OH,
otherwise protected

R'O—O

O—O

—OR

Today, many different glycosylation procedures are known and are constantly being improved and further developed. Most oligosaccharide syntheses, however, can be accomplished with the help of three main reactions, which are

(i)   the Koenigs–Knorr type reactions using glycosyl halides,
(ii)   the trichloroacetimidate method employing glycosyl trichloroacetimidates, and
(iii) the use of stable glycosides such as thioglycosides and n–pentenyl glycosides
        as the glycosyl donors.

The leaving groups for these major glycosylation reactions are listed in Table 4–1 together with the usual activating reagents for each case. Carbohydrate oxazolines are also included as they serve as donors of 2–acetamido–2–deoxy–glycosyl residues. Thioglycosides and n–pentenyl glycosides are stable compounds until they are activated by a suitable promotor after which they can act as glycosyl donors.

Table 4–1. The most common leaving groups in glycosyl donors and the activators employed.

Leaving group in glycosyl donor	Activator	Comments
L = OAc	$BF_3 \cdot Et_2O$, $SnCl_4$, TMSOTf	Not for complex oligosaccharides
L = Br	$AgCO_3$, AgOTf, $Hg(CN)_2$, $HgBr_2$	Most commonly used donor
L = Cl	AgOTf, $Hg(CN)_2$, $HgBr_2$	More stable than glycosyl bromide
L = F	$SnCl_2$–AgOTf	Can be combined with thioglycosides
L = OC(NH)CCl$_3$	$BF_3 \cdot Et_2O$, TMSOTf	Mild reaction conditions, widely used
L = SR	TfOH–NIS, DMTST, IDCP	Can also serve as an acceptor in the absence of thiophilic reagents
L = O(CH$_2$)$_3$CH=CH$_2$		Can also serve as an acceptor in the absence of bromine
oxazoline	p-TsOH, TMSOTf	Used for 1,2–*trans* glycosidation of 2–amino sugars

For the synthesis of simple glycosides peracetylated monosaccharides, such as β–penta–O–acetyl–β–D–glucopyranoside, can also be used as donors.

Synthesis of (2–bromoethyl) 2,3,4,6–tetra–*O*–acetyl–β–D–glucopyranoside.[1]

A solution of penta–*O*–acetyl–β–D–glucopyranoside (10.0 g, 25.6 mmol) and 2–bromethanol (2.17 ml, 30.7 mmol) in dry dichloromethane (100 ml) was placed in a 250 l–double necked round bottom flask placed in the dark and fitted with a dropping funnel. At 0°C BF$_3$·Et$_2$O (18.16 ml, 128 mmol) was added dropwise over a period of 25 minutes. The reaction mixture was then stirred at 0°C for 1 h and for a further 12 h at room temperature. Completion of the reaction was monitored by TLC (ethyl acetate–petrol ether, 1:1; R$_f$ = 0.6). The reaction mixture was diluted with dichloromethane (20 ml) and then poured onto ice water (100 ml) with stirring. The organic layer was separated and washed successively with water, saturated sodium bicarbonate and brine (satd. NaCl solution). The organic layer was dried over anhydrous sodium sulfate, concentrated on the rotary evaporator and the resulting residue was purified by column chromatography on silica gel using ethyl acetate–petrol ether (1:1) as eluent. The title glycoside (7.0 g, 60%) was obtained as a crystalline solid (for related NMR spectra see chapter 8).

The stereochemical result of a glycosylation reaction is dependent on whether or not the glycosidation is proceeding with neighboring group participation of the substituent at C–2. A neighboring group at C–2 will lead to a 1,2–*trans* glycoside, glycosylation without a neighboring effect will result in the formation of both, 1,2–*trans* as well as 1,2–*cis* glycosides.

4.1   1,2–*trans* Glycosidation

The involvement of ester groups at C–2 in a glycosidation reaction leads to acyloxonium ion intermediates which are formed from the oxocarbenium ion produced initially. Nucleophilic cleavage of the ring at C–1 can then only occur as a *trans*–cleavage yielding the 1,2–*trans* oriented *O*–glycosidic linkage. Nucleophilic attack of the alcohol component at the dioxolane ring carbon of the oxocarbenium ion leads to the formation of orthoesters, which might eventually be isomerized to the respective glycosides. The formation of orthoesters can become the main reaction when neutral or basic reaction conditions are being applied. In the case of axial orientation of the 2–*O*–acyl group, such as in mannose, the intermediate acetoxonium ion is especially favored due to the reverse anomeric effect, and consequently formation of orthoesters may be the predominating reaction during mannosylation. Using benzoates or pivaloates as the protecting groups in the 2–position strongly reduces the tendency for orthoester formation compared to that when acetyl groups are used; in the case of a 2–*O*–ether group, 1,2–orthoester formation is impossible.

Thus, glycosidations proceed in a straightforward manner when 1,2–*trans* glycosides are the target molecules and C–2 neighboring groups are involved. This is the case for β–glucosides, β–galactosides, and α–mannosides. On the other hand, the synthesis of 1,2–*cis* glycosides such as α–glucosides and α–galactosides is problematic and especially the synthesis of β–mannosides is a particularly well–known problem.

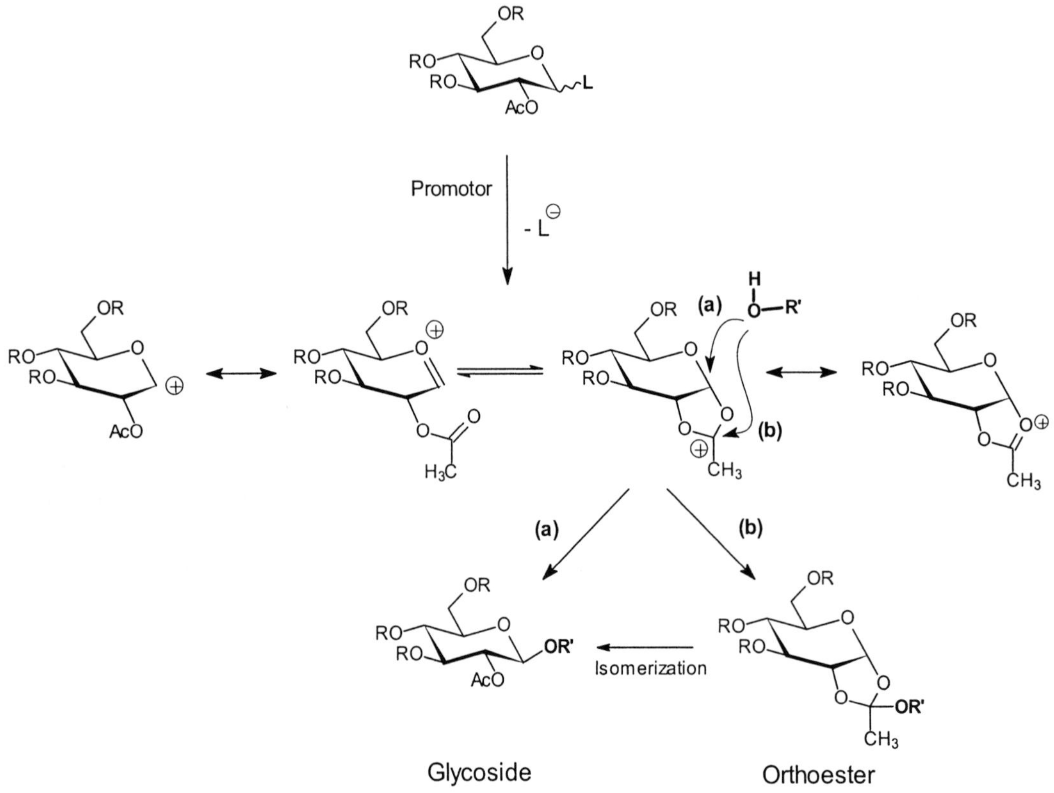

Glycosidation pathways with a participating group in the 2–position.

4.2   1,2–*cis* Glycosidation

1,2–*cis* Glycosidation is much more difficult than the synthesis of 1,2–*trans*–glycosides. Firstly, the substituent at C–2 has to be chosen as a non–neighboring group such as an O–alkyl ether. Then, an S_N2 reaction at the anomeric center of a β–glycosyl bromide would furnish α–glucosides and α–galactosides, for example. This is, however, not practical because β–pyranosyl halides, especially the bromides are greatly destabilized by the anomeric effect. A proposal which helped to solve this problem was made by R. Lemieux and co–

workers who showed that α–pyranosyl bromides react in the presence of tetraalkyl ammonium bromide with the bromide anion to produce the β–pyranosyl bromides *in situ*. The highly reactive β–pyranosyl bromide reacts much faster than its α–analog to give the α–glycoside in large proportions in a kinetically–controlled reaction. This method has been called *in situ* anomerization. It works especially well with galactose or fucose as donors, it is less effective for glucose and β–mannosides cannot be obtained at all by this approach, unfortunately. Moreover, the proportion of α–glycoside is decreased in parallel with lower donor activity and diminished reactivity of the acceptor hydroxyl group.

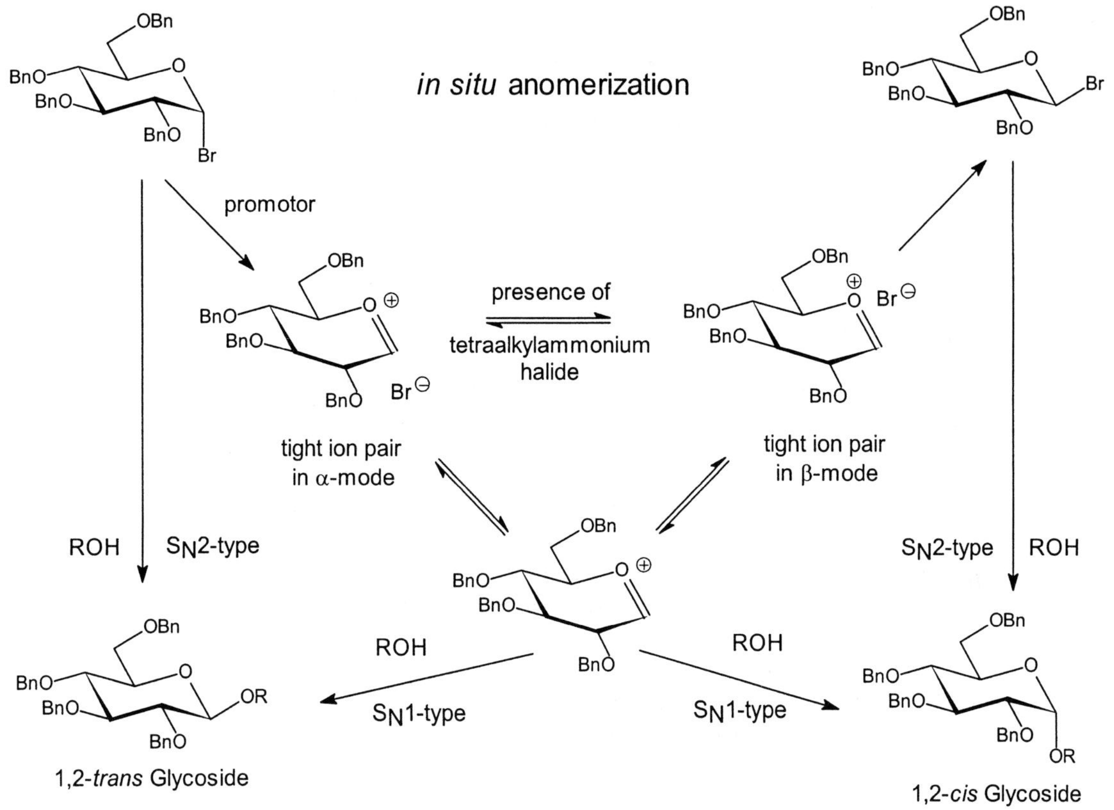

4.3   The Koenigs–Knorr method

The oldest and still the most widely–used method for the stereospecific synthesis of 1,2–*trans* glycosides is the Koenigs–Knorr reaction, introduced by Wilhelm Koenigs and Eduard Knorr from the University of Munich in 1901. The classical glycosyl donors of the Koenigs–Knorr and related reactions are glycosyl bromides and glycosyl chlorides. Silver salts serve as promotors in this reaction, both insoluble salts such as Ag_2O and Ag_2CO_3, as well as those that are soluble such as AgOTf and $AgClO_3$. A modification of this method according to Helferich employs mixtures of mercury salts, such as $HgBr_2$ and $Hg(CN)_2$. With participating protecting groups at the 2–position the Koenigs–Knorr reaction normally leads exclusively to the 1,2–*trans* glycosides, especially when insoluble silver salts are used as promotors. In this case the reaction is thought to occur at the surface of silver carbonate, assisting the formation of the *trans* glycoside by shielding the α–face of the sugar; silver bromide is formed as byproduct and precipitates out of the reaction mixture.

It may be worth reading part of the original report by Koenigs and Knorr about their glycosylation experiments as an example of a typical German publication of the time:

Synthese von Tetracetyl–β–Methylglucosid.

Berichte der Deutschen Chemischen Gesellschaft zu Berlin 34 (1901) 957–981:
Wilhelm Koenigs und Eduard Knorr: Über einige Derivate des Traubenzuckers und der Galactose. Mitteilung aus dem chemischen Laboratorium der Königlichen Akademie der Wissenschaften zu München. 1901
Tetracetyl–β–Methylglucosid, C_6H_7O (O· $C_2H_3O)_4$ · OCH_3.

8.4 g krystallisirte Acetobromoglucose wurden in 126 ccm absolutem Methylalkohol gelöst und mit 8.5 g getrocknetem, gepulvertem Silbercarbonat bei gewöhnlicher Temperatur geschüttelt. Anfangs entweicht viel Kohlensäure, wenn die Entwicklung derselben nachlässt -etwa nach einer Stunde-, bringt man die Stöpselflasche, in welcher sich die methylalkoholische Lösung der Acetobromglucose und das Silbercarbonat befinden, in eine Schüttelmaschine und setzt das Schütteln unter zeitweiligem Lüften des Stopfens so lange fort, bis eine Probe der Lösung sich als bromfrei erweist. Das war nach etwa 6 Stunden der Fall.

Jetzt wurde die Lösung vom gebildetem Bromsilber und dem überschüssigen Silbercarbonat ausgewaschen. Das methylalkoholische Filtrat, welches schwach sauer reagirte, wurde mit Wasser und einer geringen Menge reinen Baryumcarbonats versetzt, abfiltrirt, im Vacuum–Exsiccator über Schwefelsäure eingedunstet und der Trocken–Rückstand mit Wasser und Äther geschüttelt. Die vereinigten ätherischen Lösungen wurden mit Sodalösung und schließlich mit Wasser gewaschen und mittels geglühtem Natriumsulfat getrocknet. Beim Verdunsten des Äthers hinterblieb eine schön krystallisirte Substanz, welche Fehling'sche Lösung beim Kochen noch schwach reducirte. Nach wiederholtem

Umkrystallisiren aus möglichst wenig Methylalkohol zeigte die Substanz kein Reductions-vermögen mehr. Sie wurde nun noch aus hochsiedendem Ligrion umkrystallisirt. Die Aus-beute an dieser reinen Verbindung, welche das bisher unbekannte Tetracetyl–β–Methylglucosid darstellt, betrug 3.75g aus 8.4g Acetobromglucose. Das Tetracetyl–β–Methylglucosid sintert bei 102° und schmilzt bei 104–105°. Hr. Stephanowitsch hatte die Güte die Krystalle im hiesigen mineralogischen Instiut zu messen. Tetracetyl–α–Methylglucosid haben wir behufs Vergleichung mit der isomeren β–Verbindung aus α–Methylglucosid hergestellt. Das α–Methylglucosid gewannen wir nach der vortrefflichen Vorschrift von E. Fischer (diese Berichte 28, 1159 [1885]).

Synthesis of dibenzyl (2,3,4–tri–O–acetyl–L–fucosyl) phosphate.[2]

2,3,4–Tri–O–acetyl–α–L–fucosyl bromide (3.37 g, 9.54 mmol) is dissolved in a mixture of dry dichloromethane (150 ml), dry acetonitrile (150 ml) and dry diethyl ether (15 ml) and stirred with activated molecular sieve (3 Å) under an inert atmosphere for 30 minutes. Then dibenzyl phosphate (5 g) and silver carbonate (5 g) are added and the reaction mixture is stirred at room temperature in the dark for 24 h. The mixture is filtered over a celite bed, washed with dichloromethane and the filtrate evaporated *in vacuo*. Flash chromatographic purification (toluene–ethyl acetate, 4:1) gives the desired dibenzyl (β–L–fucosyl) phosphate derivative in 74% yield as a colorless syrup.

Synthesis of 3–O–(3,4,6–tri–O–acetyl–2–deoxy–2–phthalimido–β–D–glucopyrano-syl)–1,6–anhydro–2,4–di–O–benzyl–β–D–galactopyranose.[3]

The glycosyl acceptor (980 mg, 2.86 mmol) is dissolved in 22.5 ml of dry nitromethane and treated with 408 mg of collidine and 870 mg of silver triflate. The mixture is cooled to -30°C and then a solution of 1.68 g (3.37 mmol) glycosyl bromide in 10 ml of nitromethane is added dropwise. Stirring is continued for 3 h at -30°C and then the reaction mixture is allowed to warm to room temperature slowly. Dichloromethane (100 ml) is added and the mixture is filtered over a celite bed. The filtrate is subsequently washed with ice–cold water, cold 3% aq. HCl and water again, dried over MgSO₄, filtered and concentrated *in vacuo*. The residue can be crystallized from MeOH and/or purified by chromatography on silica gel to yield the desired disaccharide in 78% yield.

Koenigs–Knorr type reactions were for a long time the only available methods for glyco-sylation and can still be used effectively for numerous syntheses. These reactions, however, suffer from two disadvantages: first, the intrinsic lability of glycosyl halides, with the β–glycosyl bromides being unstable due to the anomeric effect, and second the fact that heavy metal salts in the range of equimolar amounts have to be used as promotors. On the other hand, a strong advantage of the Koenigs–Knorr method is the easy availability of the donor molecules. Glycosyl bromides are obtained in one step from the reaction of the peracety-lated sugars with HBr in acetic acid. Glycosyl chlorides are obtained from the same starting material using tin tetrachloride, for example.

Glycosyl fluorides, which can be obtained from anomerically unprotected sugars by reaction with DAST or from the peracetylated derivatives with the HF·pyridine complex, are the most modern glycosyl halides to be used as glycosyl donors. They are of considerably higher stability than all other glycosyl halides; they resist hydrolysis under basic conditions and can thus be obtained in unprotected form. Also the β–glycosyl fluorides are easily prepared under kinetically–controlled conditions. Acetylated and otherwise protected glycosyl fluorides are currently quite often used in glycosylation reactions being activated with Lewis acids.

4.4   The trichloroacetimidate method

In glycosyl trichloroacetimidates it is the anomeric oxygen atom which has been derivatized with a group that is easily removed, otherwise known as a good leaving group. This makes glycosyl trichloroacetimidates good glycosyl donors which can be activated by Lewis acid catalysts such as the borontrifluoride etherate complex (BF$_3$·Et$_2$O) or trimethylsilyl trifluo-romethanesulfonate (TMSOTf). β–Glycosyl trichloroacetimidates are more stable than the respective glycosyl bromides and can be stored at low temperature for many months. The synthesis of acetyl protected glycosyl trichloroacetimidates starts with the peracetylated sugar. Before the activation step, the anomeric hydroxyl groups has to be selectively deprotected to yield the 2,3,4,6–O–protected reducing sugar. Base–treatment (K$_2$CO$_3$, NaH, or DBU) leads to an anomeric oxyanion, which adds across the triple bond of trichloroace-tonitrile to yield the desired glycosyl trichloroacetimidates. Thus, formation of an acetylated glycosyl trichloroacetimidate starting with the free sugar involves three steps compared to two for the synthesis of a glycosyl bromide.

In addition to the acetylated form, the more active benzylated glycosyl trichloroacetimidates can be equally well synthesized, starting from the 2,3,4,6–tetra–*O*–benzylated reducing sugars. Moreover, glycosyl trichloroacetimidates can be prepared very well not only from mono– but also from oligosaccharides, carrying a variety of protecting groups. Both of the anomeric glycosyl trichloroacetimidates can be obtained in pure form, depending on the base used for deprotonation of the reducing sugar. In the case of benzylated reducing glucose, for example, the β–trichloroacetimidate is formed under kinetic control in a rapid and reversible reaction. This product anomerizes in a slow, base–catalyzed reaction, which can be speeded up with a stronger base through retroreaction and anomerization of the 1–oxide ion and renewed trichloroacetonitrile addition to form the thermodynamically more stable axial α–trichloroacetimidate.

Synthesis of a trichloroacetimidate with NaH.

The reducing glucose derivative (12.0 g, 22.2 mmol) is dissolved in dry CH_2Cl_2 (100 ml), $Cl_3C–CN$ (10 ml, 99 mmol, 4.46 eq) and NaH (50 mg, 2.08 mmol, 0.09 eq) are added and the reaction mixture is stirred at room temperature. After 15 minutes TLC indicates an α/β ratio of 1:3. For anomerization and completion of the reaction more NaH (700 mg, 29.2 mmol, 1.3 eq) is added. After 2 h of further stirring at room temperature, the mixture is filtered over a celite bed and the filtrate evaporated *in vacuo*. Flash column chromatography (petroleum ether–Et$_2$O, 3:2) gives the desired trichoroacetimidate (14.6 g, 96%) as a colorless oil, which slowly crystallizes during storage in the refrigerator.

Synthesis of a trichloroacetimidate with K_2CO_3.

To the reducing glucose derivative (3.0 g, 8.61 mmol) in 20 ml of dry CH_2Cl_2, trichloroacetonitrile (2.5 ml, 24.8 mmol, 2.9 eq) and freshly dried and powdered K_2CO_3 (2 g) are added and the reaction mixture is stirred at room temperature for 2 h. The the mixture is diluted by adding 80 ml of dry CH_2Cl_2 and K_2CO_3 is removed by centrifugation. The solution is evaporated and the residue purified by flash chromatography (CH_2Cl_2–Et$_2$O, 1:1) to obtain the β–glucosyl trichloroacetimidate (2.29 g, 54%) after crystallization from Et$_2$O–petroleum ether.

Synthesis of a trichloroacetimidate with DBU.

To a solution of the reducing mannose derivative (123 mg, 0.249 mmol) in 1 ml of dry 1,2–dichloroethane Cl₃C–CN (0.3 ml, 3 mmol, 12 eq) and DBU (10 µl, 0.07 mmol, 0.28 eq) are added successively at –5°C under argon. After stirring for 10 minutes, the mixture is directly submitted to silica gel chromatography (toluene–ethyl acetate, 6:1) to give the desired α–configured trichloroacetimidate (156 mg, 98%).

The trichloroacetimidate method has been developed into a widely–applicable method by R. R. Schmidt and co–workers and is often superior to the Koenigs–Knorr method. It has been employed in numerous oligosaccharide syntheses (cf. section 4.11). The general significance of O–glycosyl trichloroacetimidates lies in their ability to act as strong glycosyl donors under relatively mild acid catalysis. In the glycosylation step, normally glycosyl donor and acceptor are mixed in an inert solvent such as dichloromethane or acetonitrile and then the reaction is started by adding a catalytic amount of a Lewis acid. The amounts of Lewis acid employed vary from case to case. In many cases a few drops of a diluted solution of TMSOTf in dry dichloromethane is sufficient to complete the reaction, in other cases repeated additions of the catalyst are required. This might be particularly necessary when orthoesters, formed as intermediates, are to be isomerized to the respective glycosides.
In general glycosylations with glycosyl trichloroacetimidates give good yields in small scale as well as in large scale glycosylations. Ether protected trichloroacetimidates are more reactive than their ester protected counterparts, as ether groups stabilize the oxonium ion, which occurs as intermediate of the glycosylation reaction. Reactive donors are handled at low temperature. The high reactivity of glycosyl trichloroacetimidates can also lead to side reactions or even decomposition of the donor before reaction with the acceptor. When the acceptor alcohol is very unreactive the donor may rearrange to the corresponding glycosyl trichloroacetamide, which has no donor activity. Consequently, low glycoside yields are obtained. To improve yield and stereocontrol of the reaction a so–called inverse glycosylation procedure is often used, where the glycosyl acceptor and the catalyst are dissolved together first and the glycosyl donor is added then. This method often leads to improved yields because donor decomposition by the catalyst can be avoided.
Also in the trichloroacetimidate method, neighboring group participation of 2–O–acyl (or 2–N–acyl) protecting groups is usually the dominating effect in anomeric stereocontrol during glycosidation, giving rise to 1,2–*trans* glycosides. When non–participating protecting groups are selected, S$_N$2–type reactions can be carried out assisted by the use of nonpolar solvents, low reaction temperatures, and weak Lewis acid catalysts (BF₃·Et₂O). Hence, α–trichloroacetimidates yield β–glycosides and β–trichloroacetimidates α–glycosides. Strong acid catalysts (TMSOTf, TfOH), higher temperatures and more polar solvents support the formation of the thermodynamically more stable glycosylation products, which are those of the α–manno and α–gluco type.

Synthesis of benzyl 4–O–(2,3–di–O–acetyl–4,6–O–benzylidene–β–D–galactopyranosyl)–2,3,6–tri–O–benzoyl–α–D–glucopyranoside.[4]

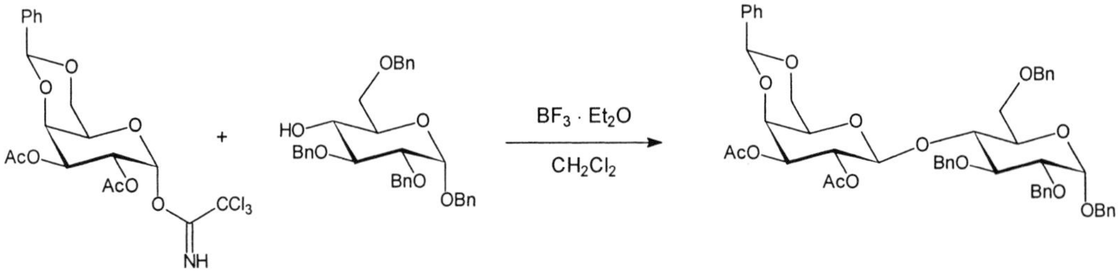

To a solution of the acceptor glycoside (10.87 g, 20 mmol) and the galactosyl tri-chloroacetimidate (15.90 g, 32 mmol, 1.6 eq) in dry CH_2Cl_2 (25 ml) under nitrogen $BF_3\cdot Et_2O$ (4.0 ml, 32.5 mmol, 1.62 eq) is added at 0°C. After stirring at room temperature for 30 min-utes, the mixture is poured into ice–cold saturated aqueous $NaHCO_3$ with vigorous stirring. The aqueous layer is extracted twice with 50 ml of Et_2O and the combined organic phases are washed with water, dried ($MgSO_4$), filtered and evaporated. Chromatography on silica gel (petroleum ether–ethyl acetate, 65:35) and crystallization from ethyl acetate–hexane gives the disaccharide (14.7 g, 84%) as colorless crystals.

Synthesis of a Galα(1,3)–Galβ(1,4)–Glc(2–deoxy–2–azido) trisaccharide, a tumor–associated antigen structure.[5]

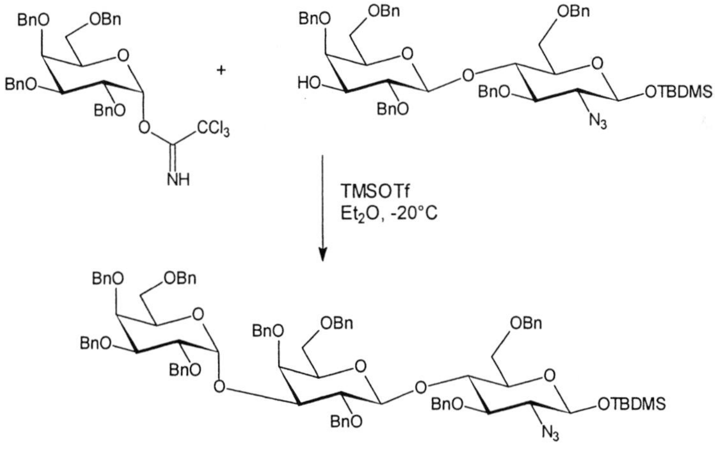

The disaccharidic glycosyl acceptor (2.06 g, 2.21 mmol) is carefully dried and dissolved in 44 ml dry Et_2O together with the galactosyl trichloroacetimidate (3.03 g, 4.42 mmol, 2 eq). The mixture is cooled to –20°C under argon and is then treated dropwise with 0.033 M TMSOTf in Et_2O (1.5 ml, 50 μmol, 0.023 eq). After 5 h, solid $NaHCO_3$ is added, the mixture is filtered and concentrated. The residue is purified by flash chromatography (petroleum ether–ethyl acetate, 8:2) to yield the trisaccharide (2.40 g, 75%) as a syrup.

Synthesis of tris[O–(2,3,4,6–tetra–O–benzyl–α–D–mannopyranosyl)oxypropyl] nitromethane.[6]

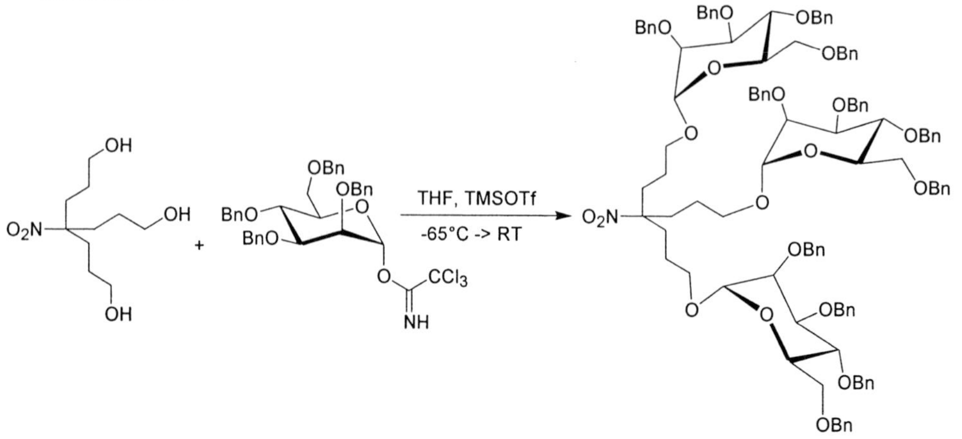

The mannosyl donor (7.48 g, 10.92 mmol) and the triol (660 mg, 2.81 mmol) are dissolved in dry THF (100 ml) under an atmosphere of nitrogen and stirred together with molecular sieves (4 Å, 4 g) at –65°C for 1 h. The reaction is started by the dropwise addition of TMS–OTf (0.02 M solution in CH_2Cl_2, 100 µl) over 10 minutes and the reaction mixture is then stirred for 2 h at –65°C and 3h at –30°C. Additional TMS–OTf (0.02 M solution in CH_2Cl_2, 150 µl) is added and the reaction mixture stirred at room temperature overnight. Then saturated aq. $NaHCO_3$ solution (100 ml) is added and the aq. phase is extracted with CH_2Cl_2 (150, 100 and 50 ml). The combined organic phases are concentrated and purified by flash chromatography (light petrolether–ethyl acetate, 3:1) to afford the benzylated trivalent cluster mannoside (3.75 g, 74%) as a colorless syrup.

4.5   Thioglycosides

Thioglycosides are quite often the glycosyl donors of choice for the synthesis of oligosaccharides, especially in glycosidation of amino sugars. A great variety of methods for the preparation of alkyl and aryl 1–thioglycosides of aldoses have been described. Commonly, fully acetylated hexopyranoses react with thiols such as thiophenol or thioethanol in the presence of Lewis acids such as $BF_3 \cdot Et_2O$ to give predominantly 1,2–*trans* products. Whereas glycosyl bromides and glycosyl trichloroacetimidates do not survive deacylation, acetylated thioglycosides can be deacetylated without degradation, so that protecting group exchange is possible.

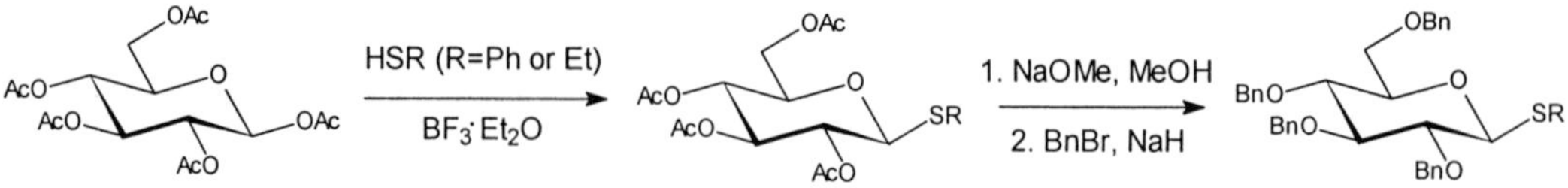

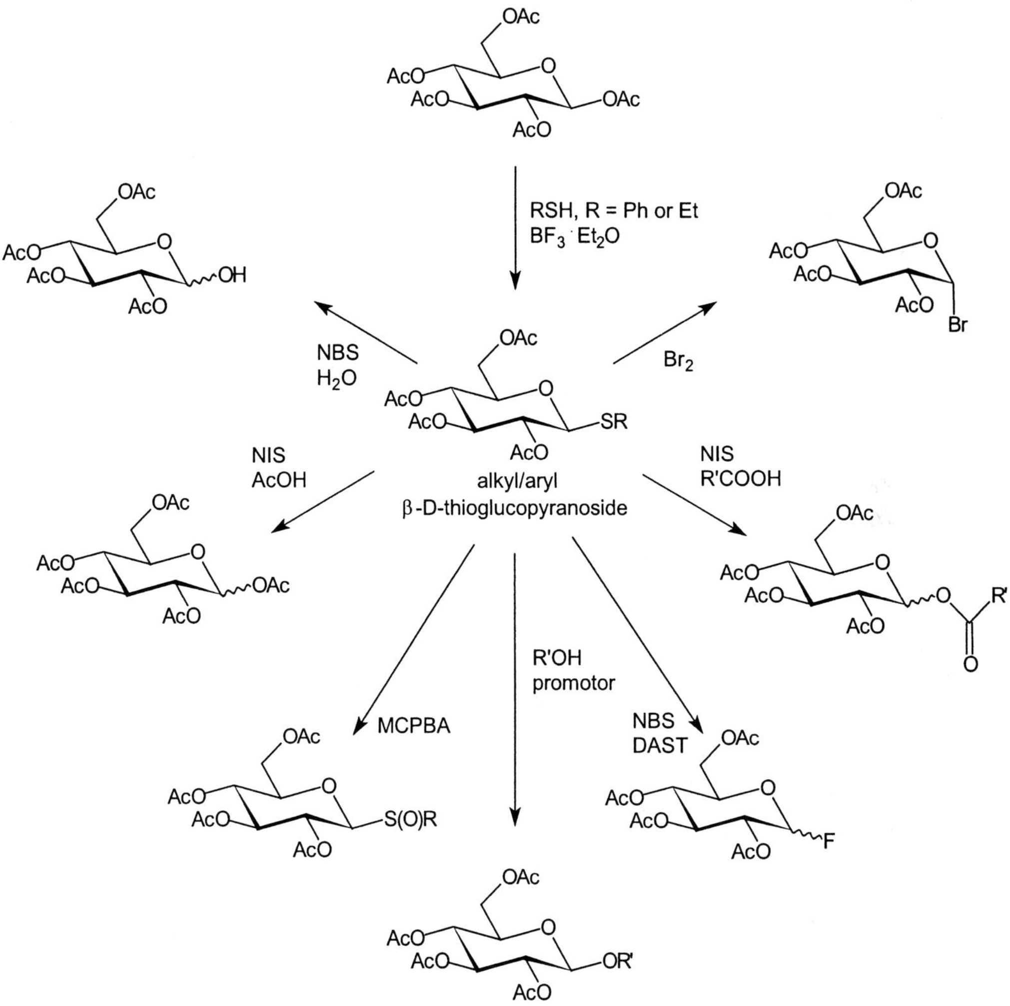

An interesting alternative for the preparation of thioglycosides is the reaction of acetylated glycosyl halides with thiourea which gives rise to a pseudothiouronium salt that can be hydrolyzed to the acetylated 1–thio–glycopyranose with aqueous potassium carbonate. This in turn can be alkylated to form the desired thioglycoside (cf. section 5.1).

Thioglycosides are stable in the absence of thiophilic promotors and interconvertable in a number of useful ways. They can be converted to all other glycosyl donors directly or by a two–step procedure. They can also be oxidized to sulfoxides, which also serve as glycosyl donors. The synthetic potential of thioglycosides is further extended by the fact that the donor activity of benzylated thioglycosides is higher than that of the acetylated analogs allowing chemoselective cross–coupling glycosidations (cf. section 4.11).

Thioglycosides can be activated by a wide variety of promotors. Often iodonium ions obtained from *N*–iodosuccinimide (NIS) or iodonium dicollidine perchlorate (ICDP), for example, are used. On the other hand, alkylating reagents can be employed in the glycosidation of thioglycosides. When the sulfur atom is methylated with trifluoromethansulfonic acid methylester (methyl triflate) the intermediate sulfonium ion which is formed can function as a good leaving group in glycoside synthesis. As methyl triflate is volatile and extremely toxic, other promotors have been suggested and proved to be effective. In particular, (dimethylmethylthiosulfonium)–trifluoromethanesulfonate (DMTST) is now regularly employed.

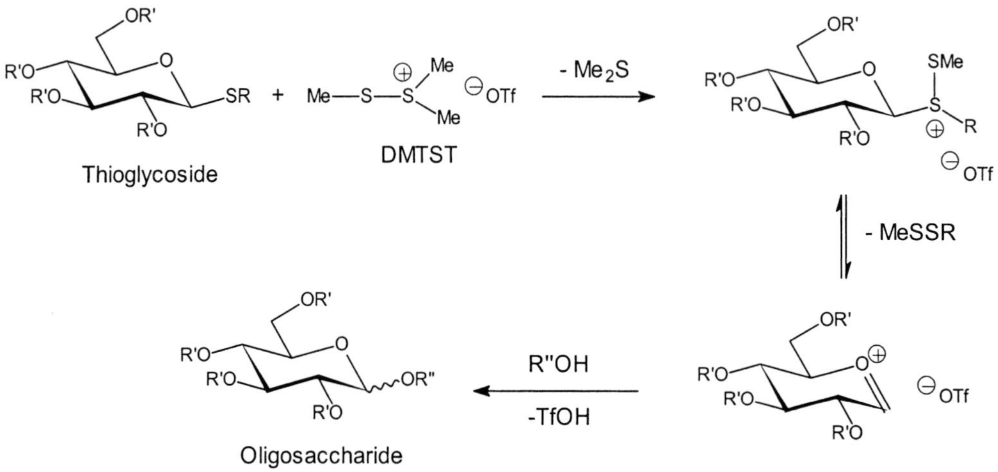

4.6   *n*–Pentenyl glycosides

4–Pentenyl glycosides are stable under most conditions except those of reductive hydrogenation. They can be prepared by Fischer glycosidation of the aldose of interest, by a Koenigs–Knorr procedure, or can be obtained from *n*–pentenyl 1,2–orthoesters. Interestingly, they can be regarded as masked glycosyl donors, as they can be activated in a sophisticated reaction introduced by B. Fraser–Reid. Promotors liberating an electrophilic halonium ion lead to the formation of a cyclic halonium ion intermediate when reacted with the pentenyl double bond. This intermediate then rearranges to a second intermediate containing the leaving group, 2–halomethyltetrahydrofurane. Elimination of this leaving group results in a glycosyl cation, which then reacts to form the glycoside.

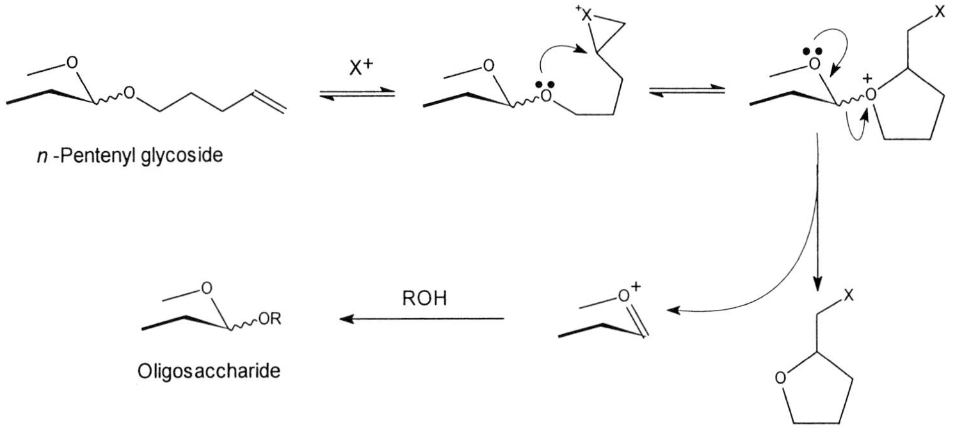

Anomeric 4–pentenoate esters have been shown to be activatable as glycosyl donors in the same way, liberating 4–iodomethyl–γ–lactone as the byproduct of the activation step.

The most widely used electrophiles for the activation of *n*–pentenyl glycosides are *N*–bromo– and *N*–iodosuccinimides (NBS, NIS) together with protic or Lewis acids such as triflic acid or triethylsilyl triflate (TESOTf). The combination NIS/TESOTf was shown to work exceedingly well. In this reaction, the activation process can be rationalized as an acid–induced heterolysis of the N–I bond.

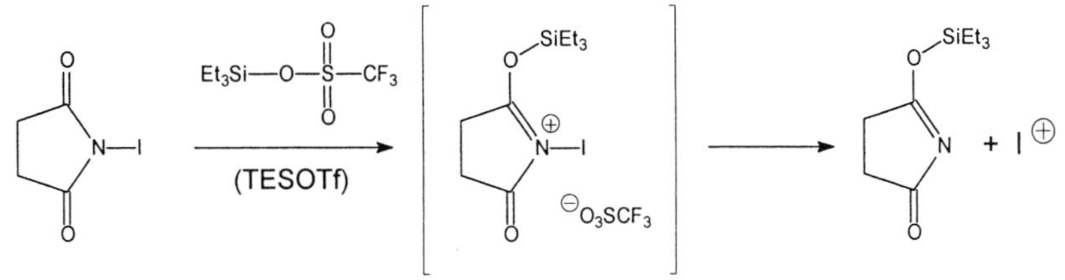

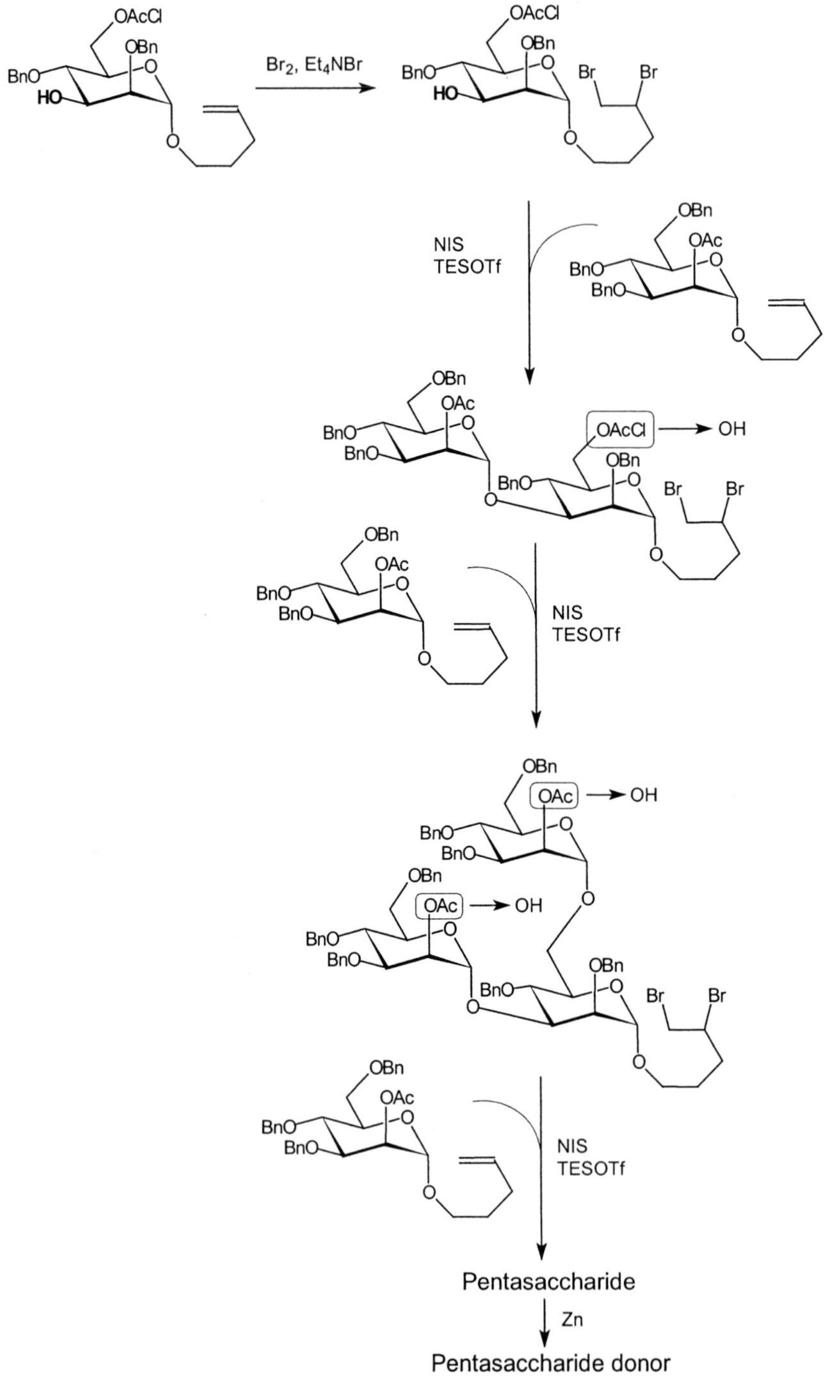

Example of the synthesis of a high mannose–type oligosaccharide using *n*–pentenyl glycosides.[7]

The activating potential of bis(2,4,6–trimethylpyridine)–iodonium–perchlorate, *N*–iodo-succinimide–trifluoromethanesulfonic acid and *N*–iodosuccinimide–trimethylsilyltriflate has been investigated; iodonium dicollidine perchlorate (IDCP) has also often been used as an activator.

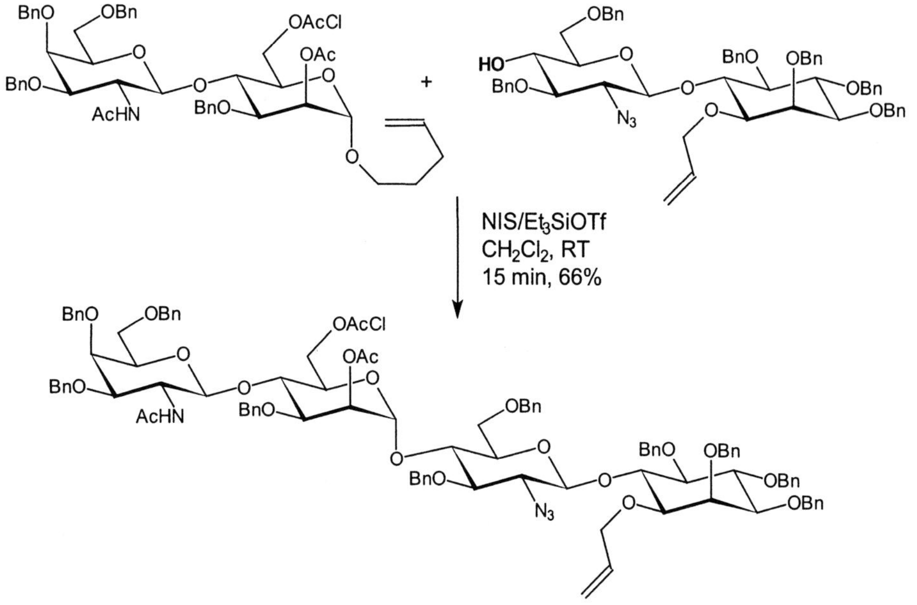

NIS/Et₃SiOTf
CH₂Cl₂, RT
15 min, 66%

n–Pentenyl glycosides are useful in the synthesis of larger oligosaccharide units. The pseudo–tetrasaccharide shown represents a segment of GPI anchor structures, consisting of a GalNAc, a mannose, a masked GlcNAc and a *myo*–inositol moiety. A set of orthogonal protecting groups (benzyl, chloroacetyl, acetyl, and allyl) have been used to allow the attachment of further saccharide building blocks at the different branch points.

The *n*–pentenyl glycosides can be converted into a number of other monosaccharide derivatives. They can be hydrolyzed to the reducing derivatives using NBS or converted into glycosyl bromides by titration with bromine solution.

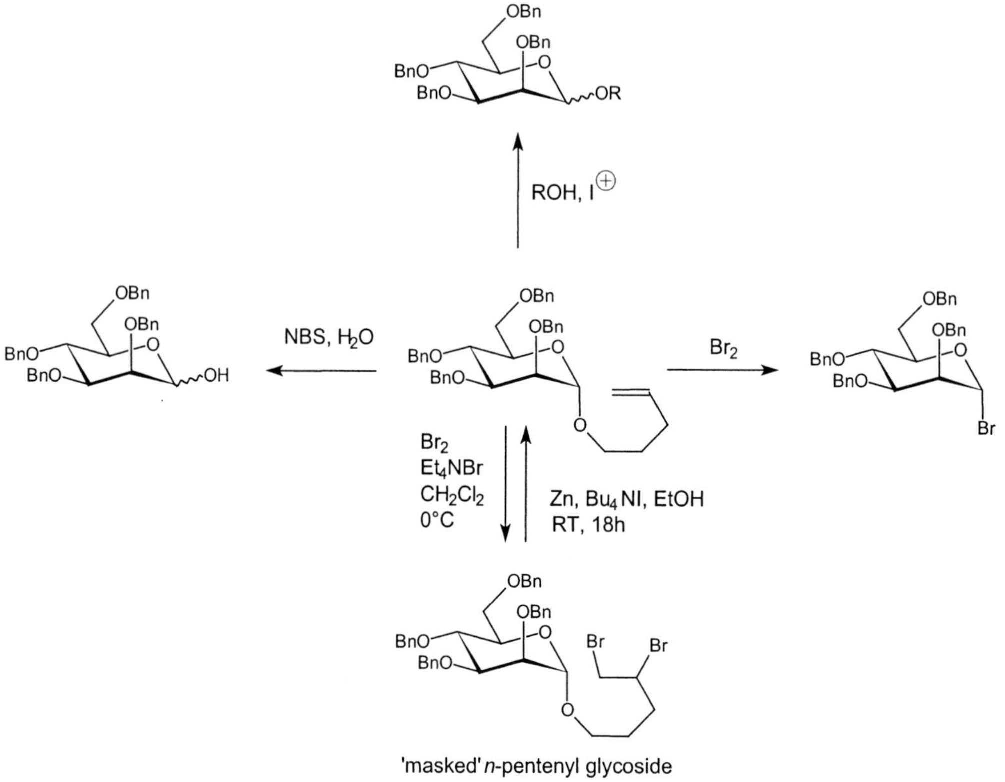

'masked' *n*-pentenyl glycoside

Of most relevance is the fact that the *n*–pentenyl moiety can become a novel protecting group, as the glycosyl donor activity can be 'switched off' by bromination of the double bond, which can later be restored using Zn dust, for example, samarium iodide (SmI₂) or sodium iodide. This possibility opens up a wide variety of synthetic options such as the consecutive assembly of high–mannose type pentasaccharide derivatives, which can be converted to pentasaccharide donors in one step by restoring the aglycone double bond using Zn, for example.

The potential of the *n*–pentenyl glycoside methodology is further increased by the observation that acyl and ether protected *n*–pentenyl glycosides display different reactivities. This finding became known as the armed–disarmed glycosylation strategy and will be discussed in section 4.11.

4.7   Synthesis of 2–deoxy–2–acetamido–glycopyranosides

2–Acetamido–2–deoxy–glucosides and –galactosides are widespread in glycoconjugate oligosaccharides in the form of α– as well as β–linked glycosides. The glycosylation proto-cols, which have to be used for their synthesis differ significantly from those applied to the synthesis of other glycosides because of the presence of the 2–acetamido group. This group has a strong tendency to form stable 1,2–oxazolines during glycosidation of 2–acetamido–glycosyl donors. These oxazolines may themselves be used as glycosyl donors, however, their use leads exclusively to the 1,2–*trans* glycosides, the synthesis of the respective α–glycosides being prohibited.[8]

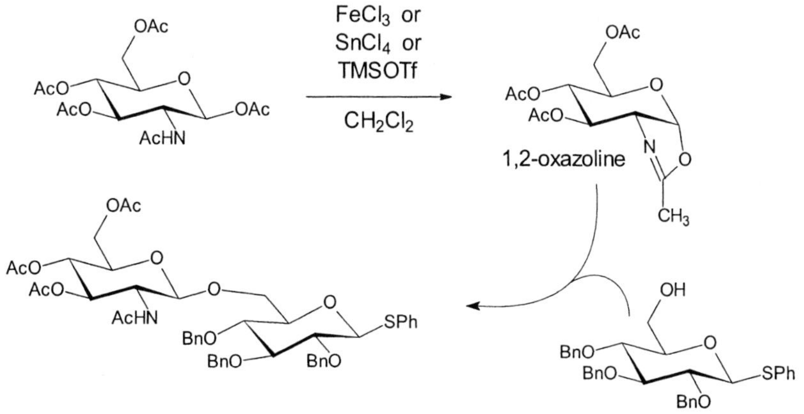

Carbohydrate 1,2–oxazolines are obtained from the respective peracetates using Lewis acids, or from glycosyl halides using heavy metal salts such as those employed in the Koe-nigs–Knorr reaction. Their activation as glycosyl donors requires rather harsh reaction conditions such as strong Lewis acids (TMSOTf, FeCl₃) or protic acids such as camphor sulfonic acid. Even then only reactive, mostly primary hydroxyl groups can be glycosylated in this reaction. To improve the donor reactivity of glycosyl oxazolines the *N*–acetyl group should be substituted with a chloroacetyl or trichloroacetyl group.

To prevent formation of 1,2–oxazolines, 2–phthalimido sugars have been extensively used and this has long been considered as the ideal *N*–protection for glycosidation of 2–amino sugars. A variety of glycosyl donors were shown to be successful as 2–phthalimido–2–deoxy–derivatives even in the glycosylation of acceptors with secondary hydroxyl groups. Like *N*–acetyl and related groups, phthalimido participates in the glycosylation step, thus allowing the synthesis of 1,2–*trans* glycosides only.

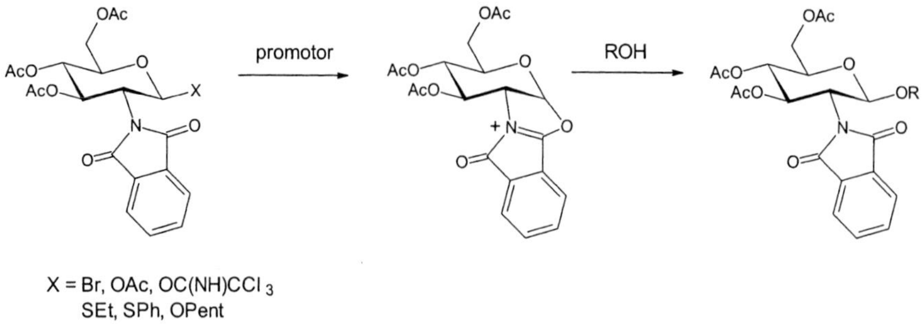

X = Br, OAc, OC(NH)CCl₃
SEt, SPh, OPent

However, use of phthalimido protected glycosyl donors is challenged by the difficulties faced in the deprotection event. The phthalimido group is cleaved with hydrazine or bases such as butylamine, hydroxylamine, or ethylenediamine under rather demanding conditions (80–100°C) and there are times when even these 'forcing' conditions fail. A series of alternatives for *N*–protection in amino sugar donors, such as dithiasuccinoyl (Dts), *N*–pentenoyl, *N,N*–diacyl protection (cf. section 3.6) have therefore been investigated. The phthalimido group itself can be activated by electron–withdrawing groups and thus the tetra-chlorophthalimido (TCP) group has been introduced with great success. Cleavage of the TCP group succeeds with two to four equivalents of ethylendiamine at 60°C, conditions which are compatible with benzoates and even acetyl groups in the molecule.

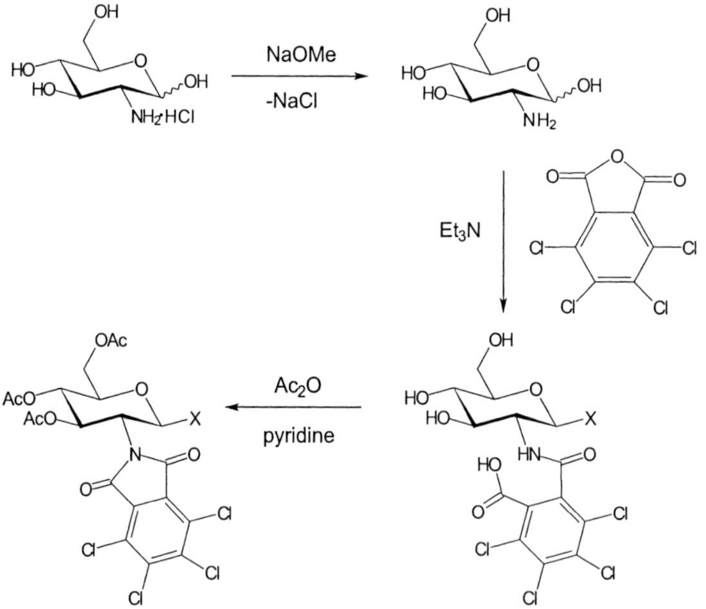

Starting from the unprotected amine hydrochloride, the free amine is liberated by treatment with NaOMe in methanol. Sodium chloride is filtered off and then one equivalent of tetrachlorophthalic anhydride is added quickly to the filtrate to avoid decomposition of the amino sugar. One equivalent of base, such as triethylamine is added, so that all of the sugar is consumed and not trapped as the ammonium salt. This produces the TCP protected amino sugar, which is soluble in methanol in the case of glucosamine. The solvent is removed *in vacuo* and then acetylation leads to the product; this is obtained via a mixed anhydride intermediate, which allows for the modestly nucleophilic amide nitrogen atom to close the imidic ring.

TCP Protection has been shown to be compatible with a variety of glycosylation methods. It, however, also leads to 1,2–*trans* glycosides, only. For the synthesis of 1,2–*cis*–2–deoxy–2–amino glycosides, use of a 2–azido moiety is favorable as this can serve as both, a latent function and a protecting group. It permits the stereoselective synthesis of α– as well as β–glycosides.

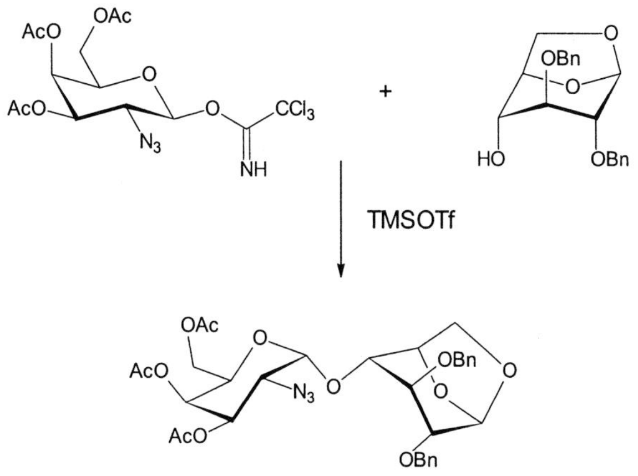

2–Azido derivatives of carbohydrates can be prepared by a variety of methods, including azido nitration of glycals (section 5.4), nucleophilic substitution of 2–sulfonates, opening of 2,3–epoxides or diazo transfer with 2–acetamido sugars. The latter method was introduced to carbohydrate chemistry by A. Vasella and employs trifluoromethanesulfonyl azide (TfN$_3$) in a reaction which can be catalyzed by copper sulfate.

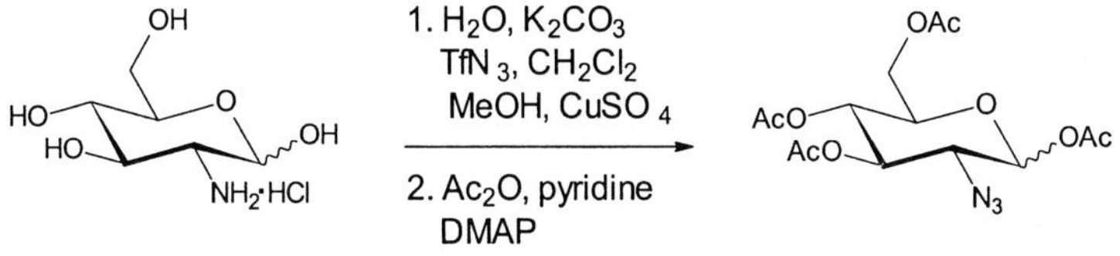

2–Deoxy–2–amino glycosides may also be obtained using the so–called Heyns rearrangement named after Kurt Heyns. In this reaction a ketose is reacted with ammonia, an amine, or an amino acid to give a ketosyl amine, which then rearranges to the respective amino sugar.

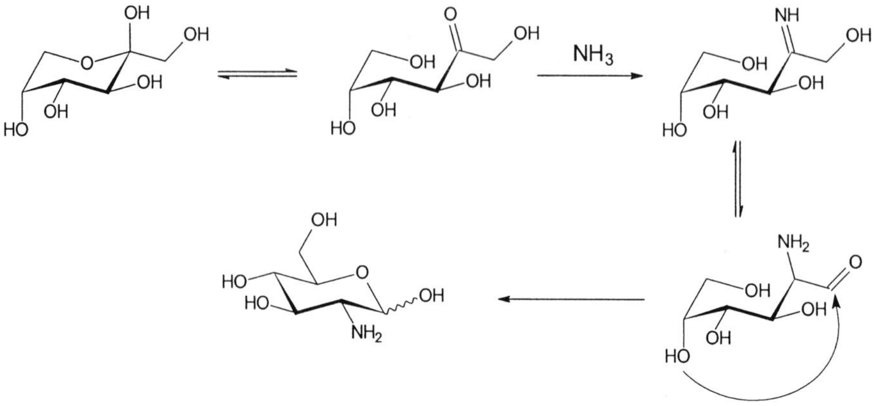

The Heyns rearrangement.

This rearrangement was used to synthesize N–acetyl–D–lactosamine from the unprotected disaccharide Gal–β–1,4–Fru in a reaction sequence which can be carried out in the same reaction vessel. Once the disaccharide ketosamine is formed, the N–benzyl group is removed by catalytic hydrogenation. Then the amino group in the glycosylamine which results of the Heyns rearrangement is subsequently acetylated in a chemoselective reaction which leaves the hydroxyl groups unprotected.[9]

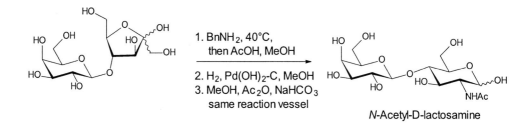

1. BnNH$_2$, 40°C,
   then AcOH, MeOH

2. H$_2$, Pd(OH)$_2$-C, MeOH
3. MeOH, Ac$_2$O, NaHCO$_3$
   same reaction vessel

N-Acetyl-D-lactosamine

4.8   Synthesis of β–mannopyranosides

The β–mannopyranosidic linkage is a common structure in glycoproteins. The chemical synthesis of this 1,2–*cis*–glycosidic linkage is, however, especially difficult. The α–mannosidic linkage is strongly favored because of the concomitant occurrence of both the α–directing anomeric effect and the repulsion between the axial C–2 substituent and the approaching nucleophile. Moreover, neighboring group participation of a 2–acyl substituent leads to α–mannosides only.

For β–mannoside synthesis the *in situ* anomerization method for the preparation of 1,2–*cis* glycosides was not successful, however, α–glycosyl bromides can sometimes be employed successfully together with certain insoluble silver catalysts such as silver oxide, silver zeolite and silver silicate.[10] The silver salts apparently direct the glycosylation mechanism toward an S_N2–type replacement so that β–mannosides are obtained by inversion of the anomeric configuration. This can be explained by shielding of the glycosyl bromide α–face provided by the interaction of silver silicate and the anomeric halide.

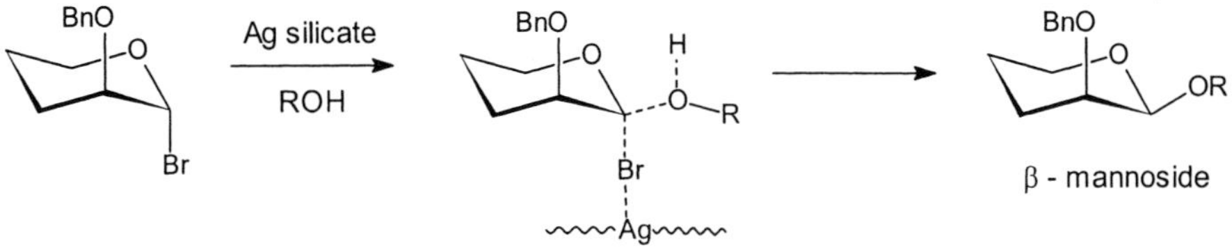

An alternative procedure for the synthesis of β–mannosides starts with the synthesis of the easily accessible 1,2–*trans*–β–glucosides followed by epimerization at C–2 in an oxidation–reduction sequence. In the reduction of the 2–oxo intermediate equatorial hydride attack rather than axial attack is favored and therefore β–mannosides can be obtained in this pathway.[11]

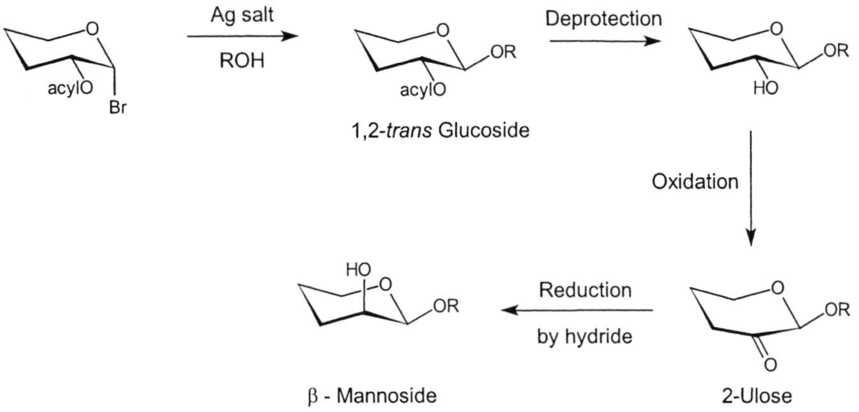

There is no absolute method for the synthesis of β–mannosides and different approaches will be taken for different cases. There is, however, a new class of methods for the preparation of β–mannosides, which is differs conceptually from the other approaches in that they proceed with complete stereocontrol due to a strategy which has been termed 'intramolecular aglycone delivery'. Here, the aglycone is attached via a temporary linkage, which may use a silicon or a carbon connection for example, to the axially positioned O–2 of a suitably protected mannosyl donor. Activation of the donor results in a concerted intramolecular delivery of the former alcohol component furnishing exclusively the β–mannosidic linkage, as shown.

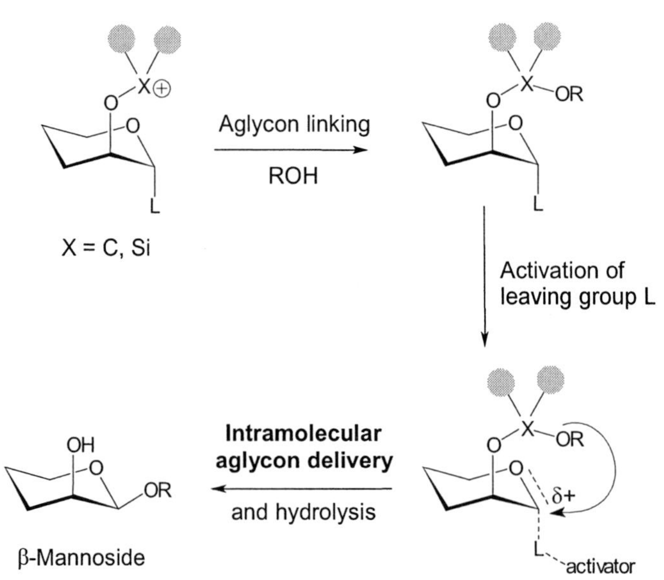

A classical example of intramolecular aglycone delivery is O. Hindsgaul's β–mannoside synthesis using a dimethyl ketal linkage at the 2–position.[12] First, a 2–O–isopropenyl ether is prepared, which can be synthesized from the corresponding 2–O–acetate by the use of a methylene transfer complex such as Tebbe's reagent, the ketal–tethered disaccharide is then formed in an acid–catalyzed reaction with an alcohol or selectively unprotected sugar derivative. This now carries the future aglycone to be intramolecularly delivered. Subsequent electrophilic activation of the aglycone–linked thioglycoside with NIS in the presence of 2,6–di–t–butyl–4–methylpyridine results in the delivery of the aglycone moiety and formation of the β–mannopyranoside.

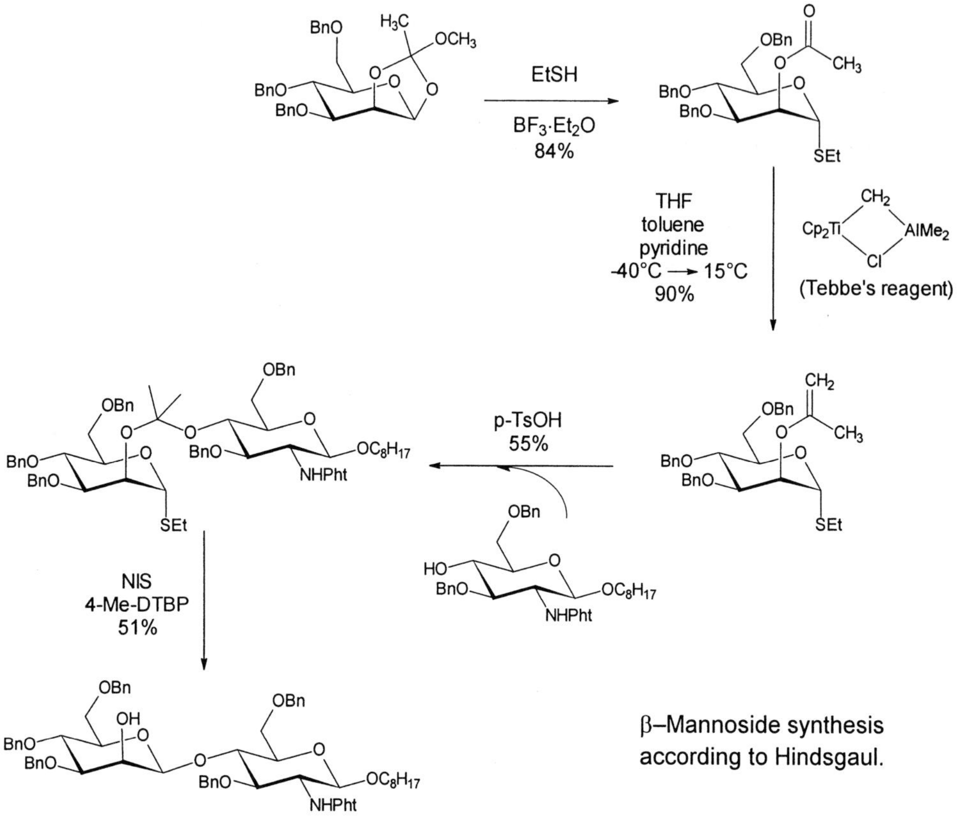

β–Mannoside synthesis
according to Hindsgaul.

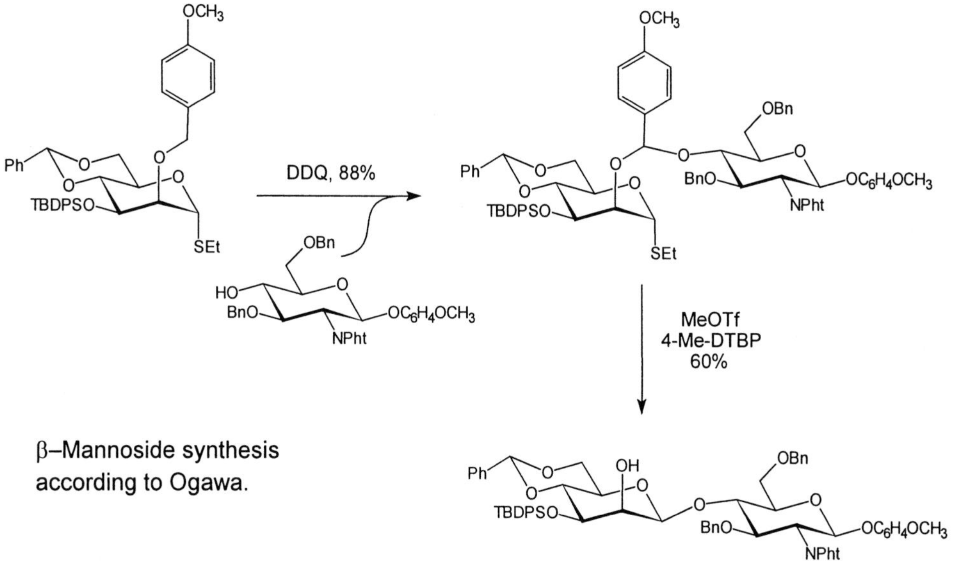

β–Mannoside synthesis
according to Ogawa.

T. Ogawa and co–workers have extended this concept to the preparation of a 2–*O*–*p*–methoxybenzyl protected glycoside, which upon activation of the *p*MBn group with DDQ in the presence of the alcohol to be glycosylated leads to a the *p*–methoxy–benzylidene acetal, prepositioning the aglycone moiety. Activation of the anomeric alkythio function, achieved by methyl triflate and 4–Me–DTBP, affords the β–mannopyranoside.[13]

In general it can be observed in modern carbohydrate chemistry that, for solving the most difficult glycosylation problems, sophisticated strategies have been successfully designed, which, however, often require complex preparation of the starting materials for the actual glycosylation reaction. This is also the case with intramolecular aglycone delivery. Recently, a direct β–mannosylation reaction has been introduced by D. Crich[14], as a variation on a glycosylation method introduced by Kahne.[15] It involves 4,6–*O*–benzylidene-protected mannosyl sulfoxides which are activated with trifluoromethanesulfonic anhydride at low temperature to form α–mannosyl triflates. These triflates then react S_N2-like with alcohols under selective formation of the β–mannoside. The reaction is typically conducted in the presence of a hindered base such as 2,6–di–*tert*–butyl–4–methylpyridine. The use of a 4,6–*O*–benzylidene protective group is required for high selectivity, as is the use of non–participating protecting groups on *O*–2 and *O*–3 of the donor molecule. A hypothesis about the mechanism of the reaction argues with solvent–separated and contact ion pairs to understand the high β–selectivity, which is observed in this mannosylation reaction.

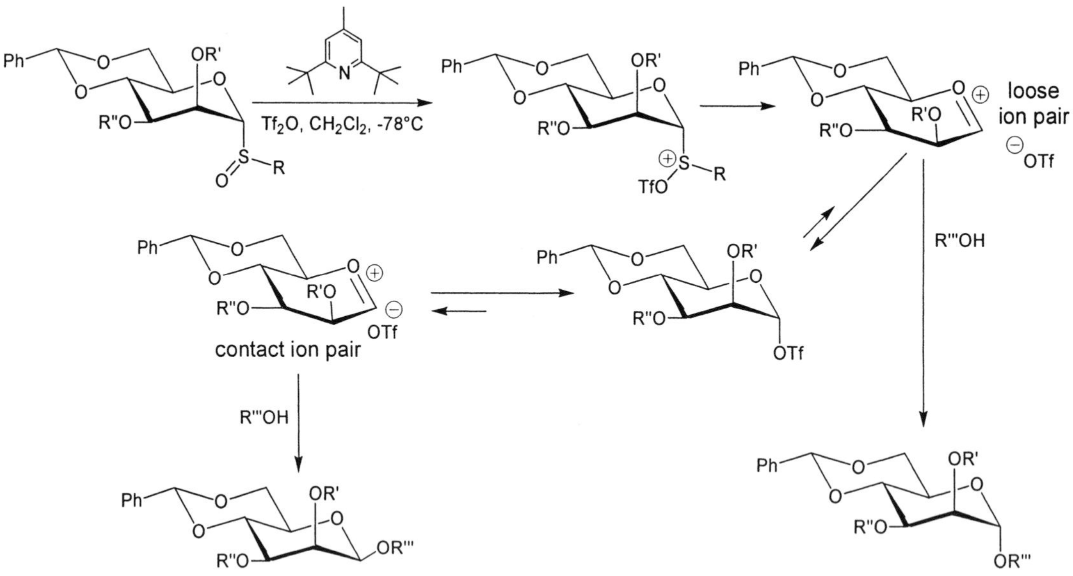

4.9   Synthesis of 2–deoxy glycosides

Deoxygenated sugars are common constituents of many biologically active glycosylated natural products such as macrolides, anthracyclines, cardiac glycosides and aureolic acids. Frequently 2,6–dideoxy glycosides are found in these cases (cf. chapter 5). Therefore, the synthesis of the respective glycosides is an important task which is however, made difficult by the fact, that in 2–deoxy glycosides no neighboring group adjacent to the anomeric center can induce stereoselectivity in the glycosylation step.

A reliable approach for the synthesis of 2–deoxy–α–glycosides is based on the use of a temporary directional functionality at C–2. J. Thiem and co–workers have shown that addition of NIS across the double bond of glycals leads to the formation of cyclic iodonium intermediates in which, due to stabilization by the reverse anomeric effect, the β–face of the carbohydrate ring is sterically shielded.[16] The stereoisomeric iodonium ion in which the α–face of the carbohydrate ring is shielded is normally not formed. Nucleophilic cleavage of the iodonium intermediate by an alcohol then consequently leads to the 1,2–*trans* 2–deoxy–2–iodo–α–glycoside, in which reduction of the iodo function leads to the 2–deoxy–α–glycoside.

The fact that the iodo function is reduced not only by catalytic hydrogenation, but also by radical reduction is advantageous, as radical reduction leaves azide functions intact.

Another approach used by Thiem's group is to add *O,O*–dialkyl dithiophosphates across the double bond of a glycal to form *S*–(2–deoxy–glycosyl)–phosphorodithioates in nearly quantitative yield.[17] These addition products are stable derivatives which can be utilized as 2–deoxy–glycosyl donors without purification. Using iodonium–(di–*sym*–collidine) perchlorate as a promotor, glycosylation of weak nucleophiles can also be effected in high yields and with high α–stereoselectivity. This has been exemplified with an L–rhamnal derivative, which, after addition of *O,O*–diethyl dithiophosphate, was glycosidated with a benzyl β–L–digitoxide derivative to yield the α–configured disaccharide, a fragment of the macrolide antibiotic kijanimicin.

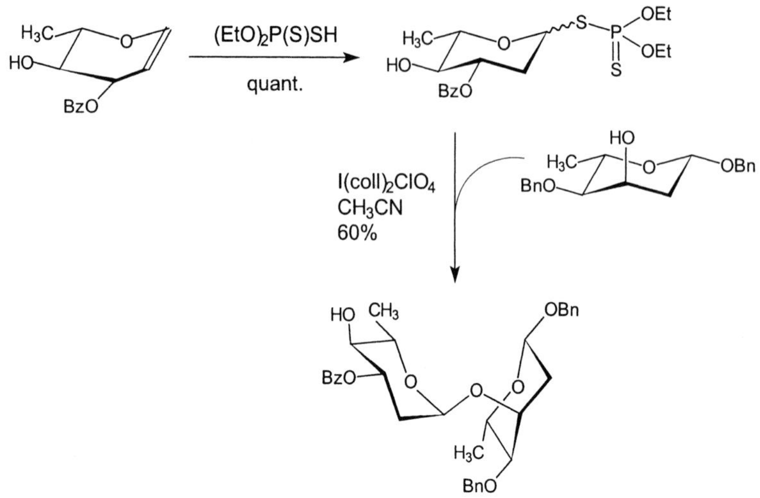

The β–2–deoxy–glycosides can be obtained by different methods. One possibility is to treat 2–OH–unprotected thioglycosides with DAST.

This reaction results in a stereoselective 1,2–migration of the fluorine substituent giving rise to 2–thiophenyl glycosyl fluorides which are efficient glycosyl donors. Depending on the solvent used, they can be converted diastereoselectively to 2–thiophenyl–α– or β–glycosides after activation with tin(IV) chloride.[18] The 2–thiophenyl group can be later removed by reduction with Raney–Ni to obtain the desired deoxy function in the glycoside.

2–Deoxy–β–glycosides can be also obtained when the so–called DBE method is used. In this method suitably protected rhamnose derivatives are treated with dibromomethyl methylether (DBE) and zinc bromide to furnish a 2–bromo–2–deoxy–glycosyl bromide which carries a formyl group at the 3–position. The so obtained glycosyl bromide can be activated with silver salts to yield 2–bromo–2–deoxy–glycosides mainly in the β–form.

In the example given, the anomeric mixture obtained was separated after radical reduction of the 2″–bromo substituent to establish the deoxy function and chemoselective cleavage of the formyl group in the 3″–position.

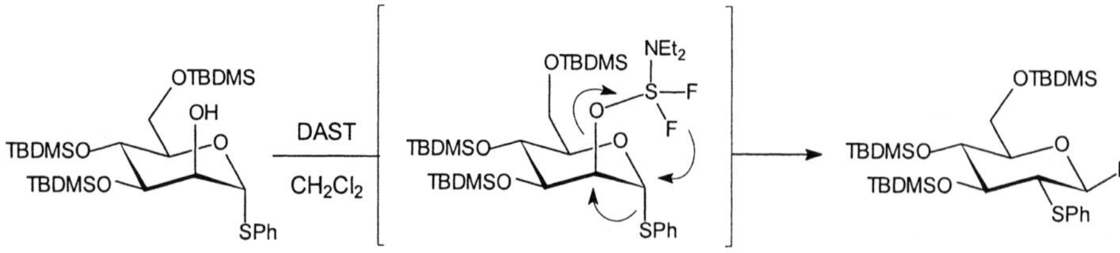

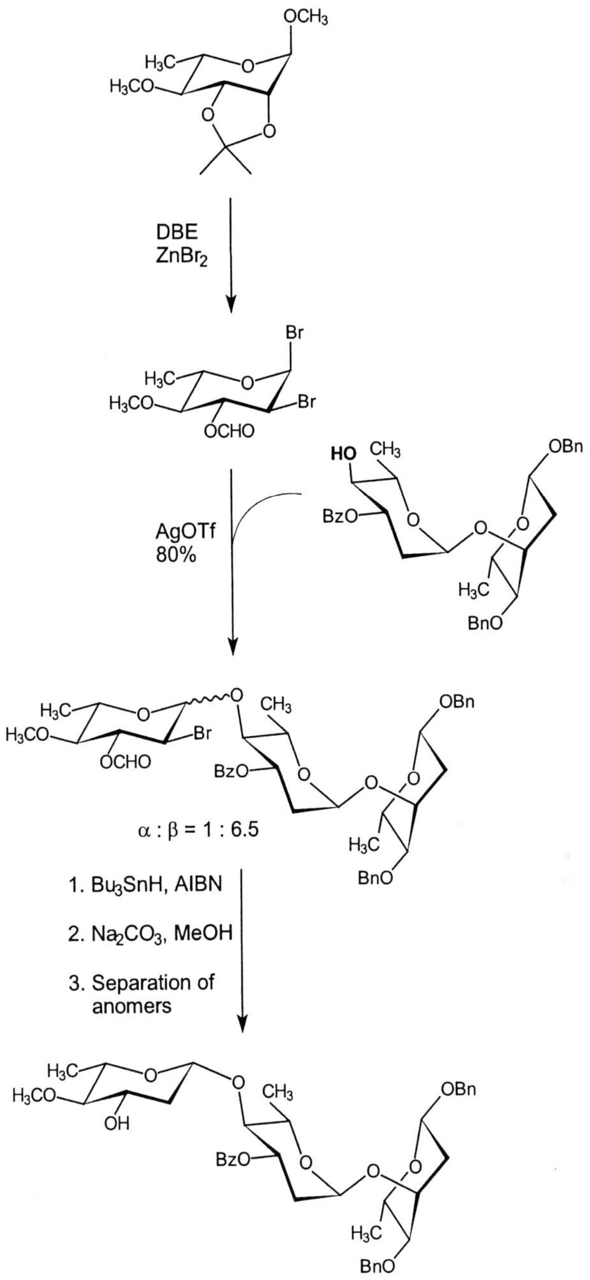

β–2–Deoxy–glycoside synthesis using the DBE method.

4.10   Synthesis of glycopeptides

To prepare partial structures present in glycoproteins, the synthesis of both *O*–linked and *N*–linked glycopeptides is of interest (cf. chapter 6). For synthesis of *O*–linked glycopeptides, glycosylated serine or threonine building blocks can be prepared using one of several standard glycosylation procedures, including the Koenigs–Knorr and the trichloroacetimidate method or by employing glycosyl fluorides or thioglycosides. For the preparation of *N*–linked glycosyl amino acids, asparagine is most often selected for the peptide coupling reaction with the desired glycosyl amine.

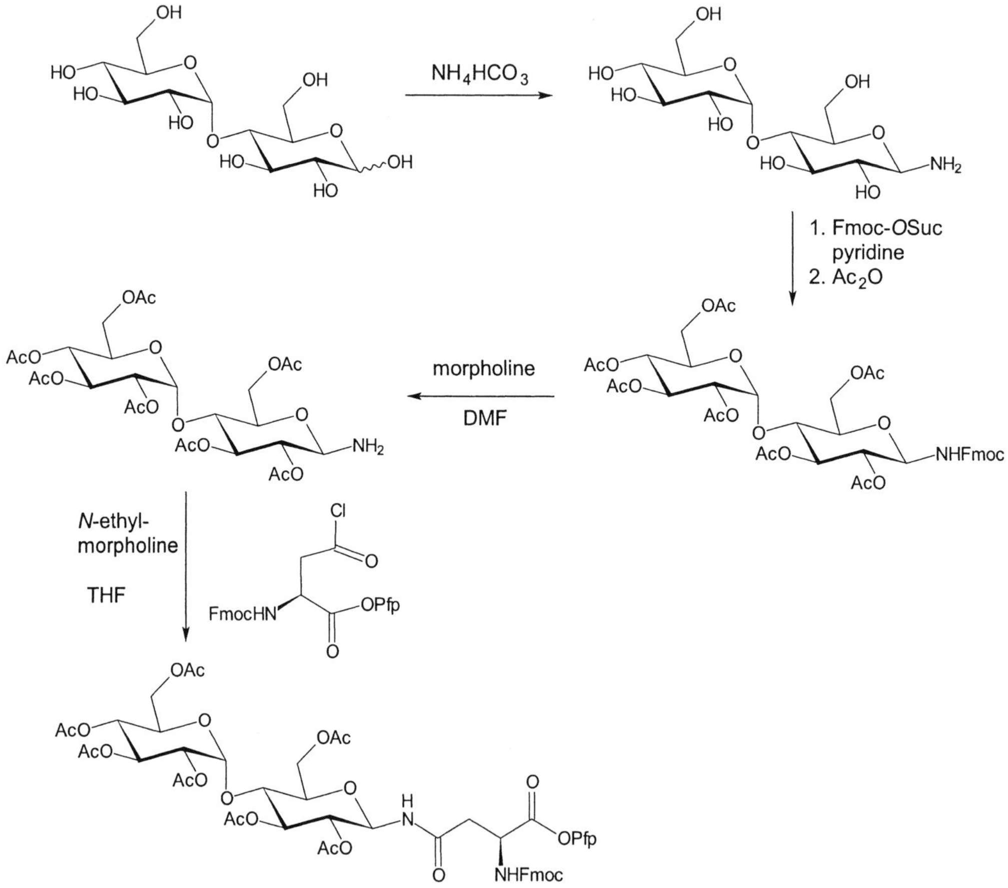

Glycosyl amines can be prepared by reduction of glycosyl azides, from 1,2–oxazoline sugar derivatives with trimethylsilyl azide or by the reaction of reducing sugars with ammonium–hydrogen carbonate; the latter approach is depicted for the disaccharide maltose. The resulting maltosylamine has to be *N*–protected prior to acetylation of the hydroxyl groups. For protection of the amino function, the fluoren–9–yl–methoxycarbonyl (Fmoc) group is frequently used. It can be removed under mild basic conditions, using morpholine, piperidine or DBU without the occurence of undesirable side reactions. After *O*–acetylation and removal of the Fmoc group, the peptide coupling reaction produces an *N*–linked malto-syl–asparagine derivative.

In order to couple a carbohydrate with an amino acid the latter has to be suitably pro-tected at the α–amino and the carboxy group. The required protecting groups are often se-lected to facilitate the eventual assembly of glycosyl amino acid building blocks to form glycopeptides. In principal, two different strategies can be considered for glycopeptide syn-thesis. Either carbohydrates are attached to the completed peptide, or the glycopeptide is obtained by peptide coupling of glycosylated amino acids. The latter approach is currently most commonly applied and often is combined with solid phase methodology.[19]

For solid phase synthesis of glycopeptides it has been shown to be advantageous to use the pentafluorophenyl (Pfp) ester as the *C*–terminal protecting group.[20] This ester is stable in both *O*– and *N*–glycosylations and also during chromatographic purification. Further-more, the Pfp ester activates the carboxylic acid for a nucleophilic attack by an amine dur-ing peptide bond formation.

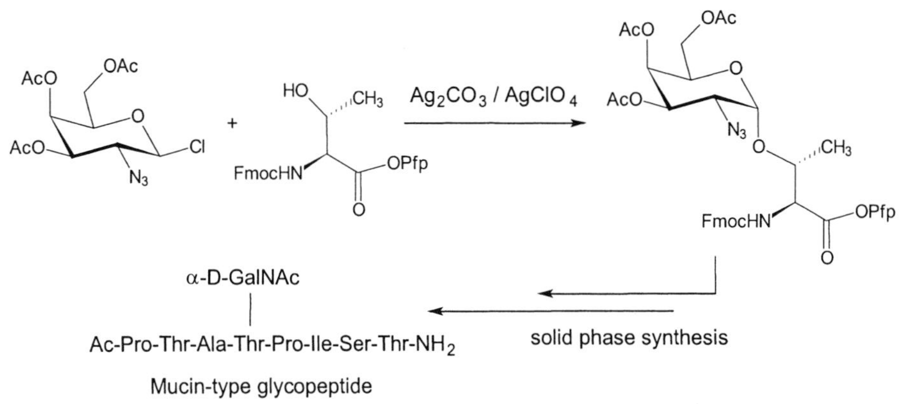

Preparation of an *O*–glycosylated threonine derivative which has been assembled to a se-ries of glycopeptides of the mucin–type by solid phase synthesis.

4.11   Glycosylation strategies

Apart from the methods for glycosylations presented herein, the most commonly used of which are the use of glycosyl halides, glycosyl trichloroacetimidates or thioglycosides, several other methods have been developed. Glycals or the 1,2–epoxides derived therefrom, the so–called Brigl anhydrides, can be used as glycosyl donors, as can glycosyl dialkyl-phosphites, glycosyl dithiocarbonates, glycosyl sulfoxides, and anomeric diazirines, to name but a few. One or the other of these approaches may be the method of choice, when particularly difficult glycosylation problem is encountered. Finally, there is one important truth in glycoside synthesis, commented on by H. Paulsen in 1982: 'Each oligosaccharide synthesis remains an independent problem whose resolution requires considerable systematic research and a good deal of know how. There are no universal reaction conditions for oligosaccharide synthesis.'[21]

Indeed the number of factors which have to be taken into consideration in oligosaccharide synthesis are immense. First, the appropriate glycosyl donor and activation reagent have to be chosen. Second, in order to differentiate between the multitude of hydroxyl groups, extensive protection strategies are required, especially for the glycosyl acceptor, only one hydroxyl group of which should normally be glycosylated. By the choice of protecting groups the donor activity as well as the stereochemical outcome of the glycosidation step is determined as shown with neighboring substituents in the 2–position resulting in the stereospecific formation of 1,2–*trans* glycosides.

Synthesis of a Lewis^X dimer intermediate.[22]

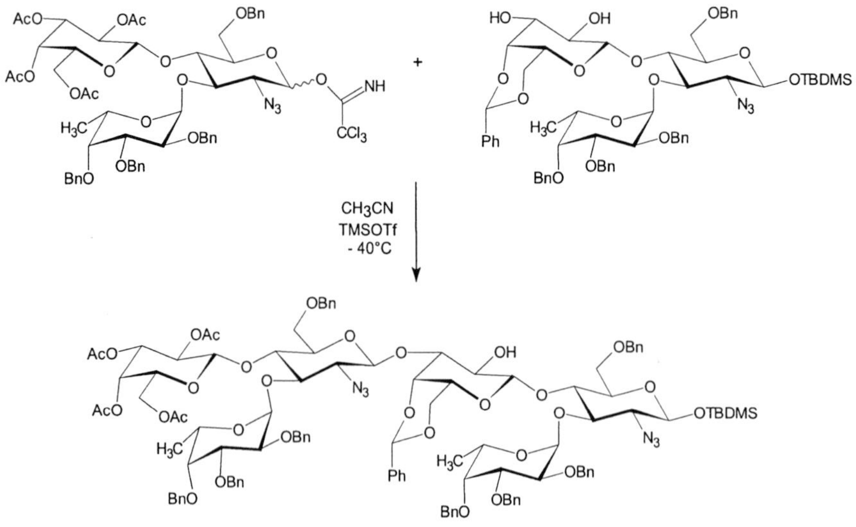

The trichloroacetimidate (4.75 g, 4.0 mmol) and the acceptor (4.84 g, 4.5 mmol, 1.12 eq) are dissolved in 60 ml dry acetonitrile and cooled to –40°C under argon. The mixture is treated with a 0.05 M solution of TMSOTf (0.8 ml, 40 µmol, 0.01 eq) and solid NaHCO$_3$ (0.5 g) is added after 10 minutes. After filtration the reaction mixture is concentrated *in vacuo*. Short column chromatography of the residue (petrol ether–ethyl acetate, 3:2) gives a color-less foam which can be crystallized from Et$_2$O–petroleum ether to give the hexasaccharide (6.72 g) in 80% yield.

Block synthesis

Besides the problems of regio- and stereoselectivity, economic strategies have to be selected particularly for the synthesis of a larger oligosaccharides. In these cases a stepwise assembly of the target molecules by consecutive attachment of one monosaccharide donor after an-other onto the growing saccharide is less advantageous than using a block synthesis which follows a convergent strategy.

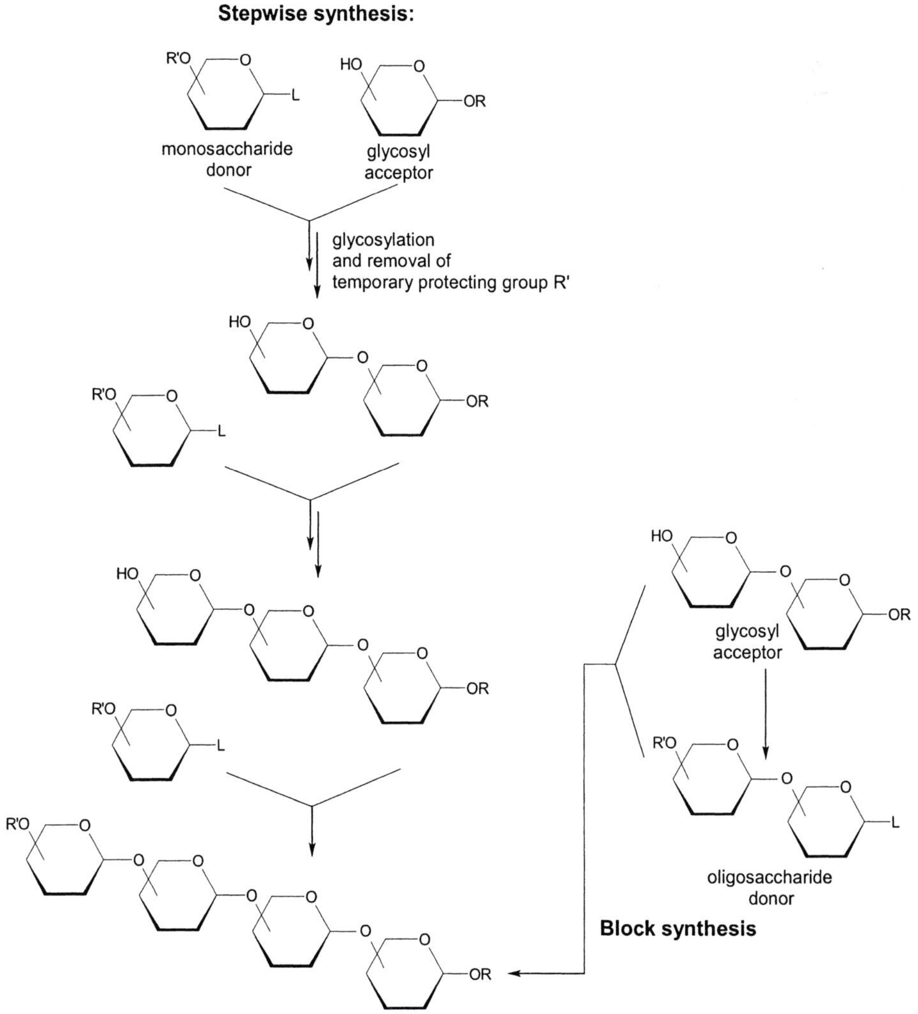

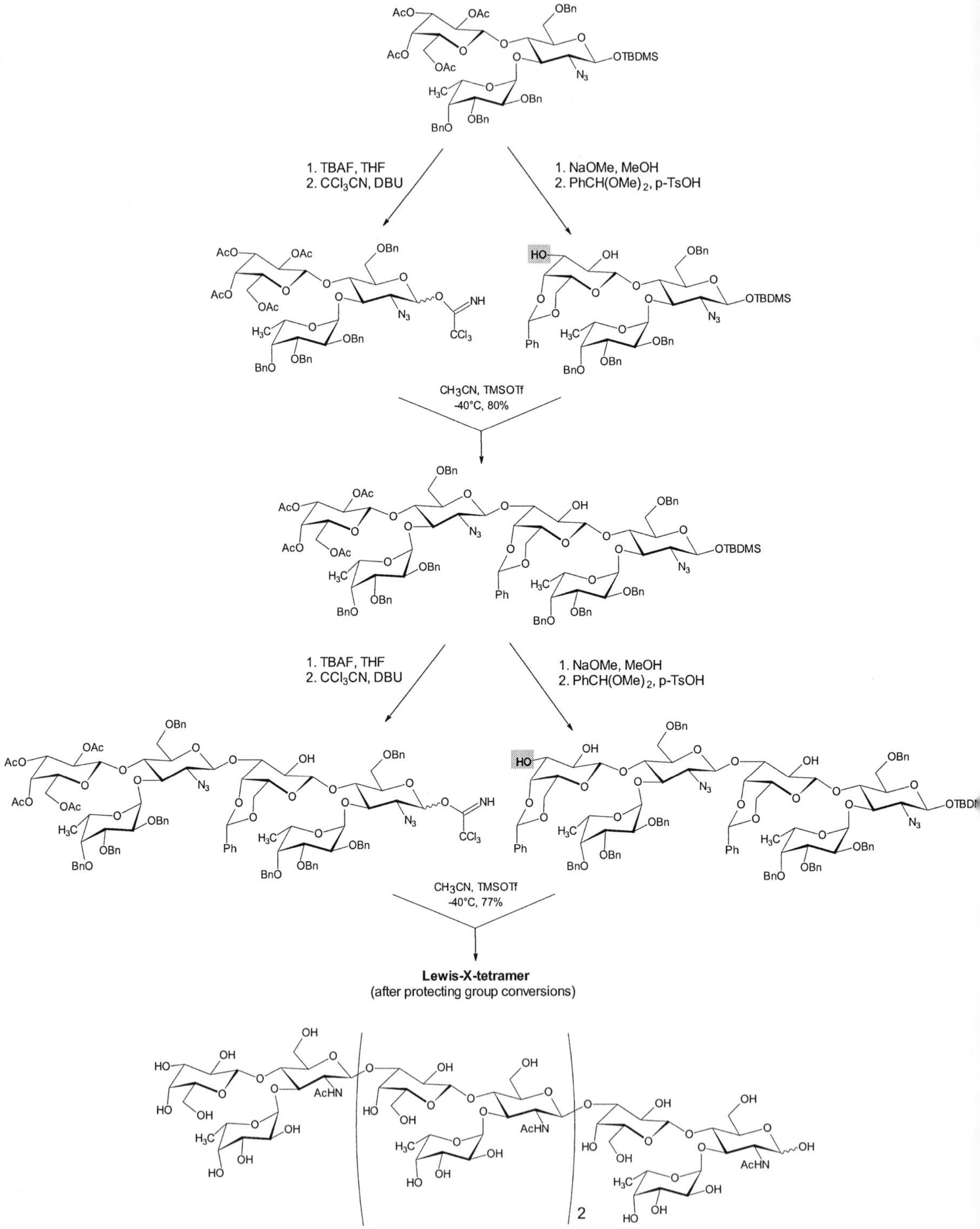

CH3CN, TMSOTf
-40°C, 80%

1. TBAF, THF
2. CCl3CN, DBU

1. NaOMe, MeOH
2. PhCH(OMe)2, p-TsOH

1. TBAF, THF
2. CCl3CN, DBU

1. NaOMe, MeOH
2. PhCH(OMe)2, p-TsOH

CH3CN, TMSOTf
-40°C, 77%

Lewis-X-tetramer
(after protecting group conversions)

Block synthesis has relatively early become a principle method in oligosaccharide synthesis and has been impressively demonstrated for the synthesis of the Lewis[X] tetramer using the trichloroacetimidate strategy. The synthesis of this dodecamer was even more economic as the 2– and 3–hydroxyl groups in the respective glycosyl acceptors did not have to be distinguished by protection but could be regioselectively glycosylated only affording the 1→3–glycosidic linkage.

The armed–disarmed concept

The different reactivities of glycosyl donors can also be strategically utilized to facilitate oligosaccharide synthesis. The reactivity of the anomeric center depends on the configuration of the saccharide unit and to a larger degree on the substitution pattern. The fundamental work on this problem was published by Paulsen's group in the late 1970s. It was observed that in general acyl groups such as acetyl groups reduce the reactivity at the anomeric center, while ether groups such as benzyl ethers increase it.

Table 4–2. The armed–disarmed concept.

Disarmed	Armed
Acyl protecting groups	Ether protecting groups
Depressed anomeric reactivity	Higher anomeric reactivity
Less activatable	Easily activatable

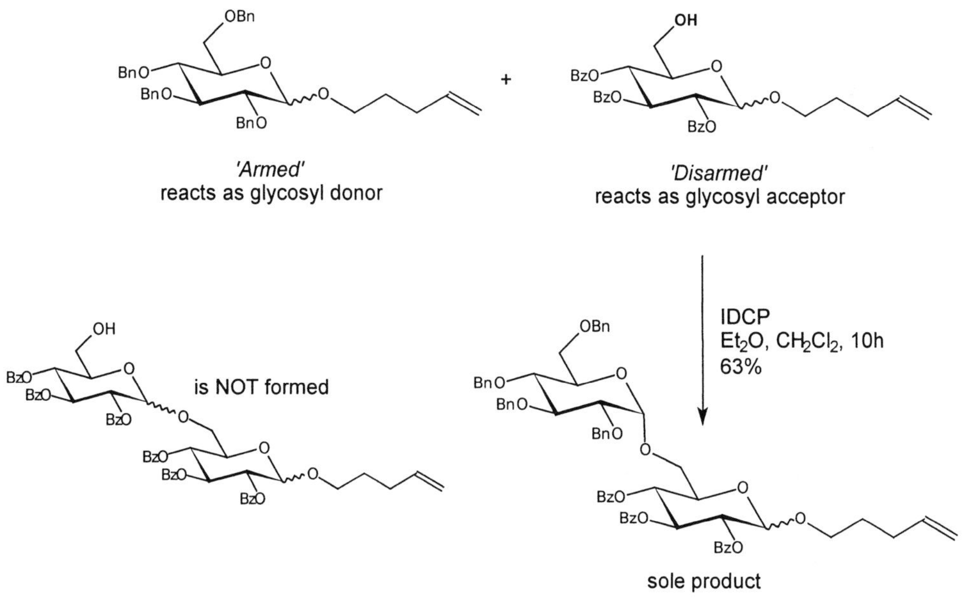

'Armed'
reacts as glycosyl donor

'Disarmed'
reacts as glycosyl acceptor

is NOT formed

IDCP
Et$_2$O, CH$_2$Cl$_2$, 10h
63%

sole product

The differences in reactivity, which may depend in particular on the nature of the C–2 protecting group, formed the basis of the so–called 'armed–disarmed' principle introduced by B. Fraser–Reid for *n*–pentenyl glycosides.[23] It was found that oxidative hydrolysis (NBS, H$_2$O) of *n*–pentenyl glycosides required minutes when the C–2 protecting group was an ether, but hours when it was an ester. This notion opened the attractive possibility of condensing an armed 4–pentenyl donor with a partially protected disarmed 4–pentenyl acceptor to obtain the cross–coupled disaccharide only, with none of the self–coupled product being formed.

The dependence of reactivity on the substitution pattern also allows thioglycosides to be combined with oligosaccharides following the 'armed–disarmed' principle.

Thus, activation of an armed, benzyl protected, thioglycoside with iodonium ions, derived from iodonium dicollidine perchlorate (IDCP) allows the glycosylation of a partially protected disarmed thioglycoside, giving only one disaccharide with the properties of the disarmed glycosyl donor. This in turn can be converted to an armed glycoside by protecting group exchange and can subsequently be submitted to the next, analogous glycosylation step with a disarmed acceptor, yielding a

disarmed trisaccharide. Thus, this sequence has an iterative character which leads, in theory, to infinite growth of linear oligosaccharides. The concept of graded reactivity of glycosyl donors is an attractive one and may be even be extended to the combination of glycosyl donors with more than two types of reactivity.

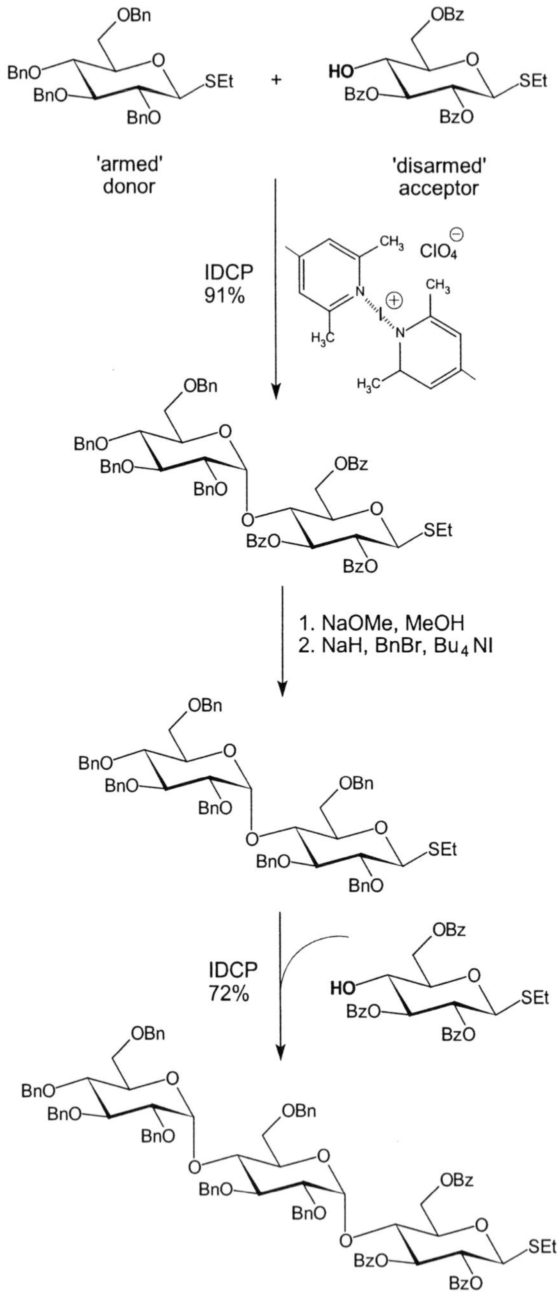

4.12 Enzymatic glycosylation

An alternative to the chemical synthesis of oligosaccharides is to employ the 'natural machinery' and use carbohydrate–specific enzymes for the glycosylation step, as well as for
the synthesis of the required substrates. Enzymatic glycosylation has the advantage of not
requiring protecting groups and of regio– and stereoselectivity being normally characteristic
of the particular enzyme used. This, consequently, also implies a restriction of this methodology in the preparation of non–natural oligosaccharide analogs for which no suitable enzymes are available. Using enzymes also has other limitations, such as restricted access to
expensive or cloned enzymes and the required co–factors, as well as tedious aqueous work–
up procedures for the removal of buffer salts for example, procedures which are not required in classical synthesis. Nevertheless, enzymatic glycosylation methodology has been
intensively developed recently and applied with great success in many cases.

Two classes of enzymes can be employed in enzymatic oligosaccharide synthesis,

(i)   the glycoside cleaving *glycosidases*, and
(ii)   the linking *glycosyltransferases*.

The readily available and easy to handle glycosidases are used in a reverse hydrolysis reaction by shifting the equilibrium of glycoside cleavage in the opposite direction. Consequently, the yields obtained from glycosidase–catalyzed glycoside synthesis are typically
low. Alternatively, glycosidases can be used for synthesis in transglycosylation–type reactions employing glycosides with good leaving groups, such as glycosyl fluorides or *p*–
nitrophenyl glycosides as glycosyl donors, as shown for *p*–nitrophenyl α–L–fucoside.[24]
These reactions can even be carried out in the presence of organic solvents which are added
to the aqueous solutions.

Glycosidases do not necessarily have to be isolated in pure form, but may be efficiently
used as crude extracts and even whole cells can be employed in certain situations. Glycosidases are also relatively tolerant of their substrates and not strictly regioselective in the saccharide linking step, so that the scope of possible products is increased compared to the use
of glycosyltransferases. Glycosyltransferases on the other hand, are highly specific regarding their donor as well as acceptor substrates, which they link together in a stereo– and regiospecific manner.

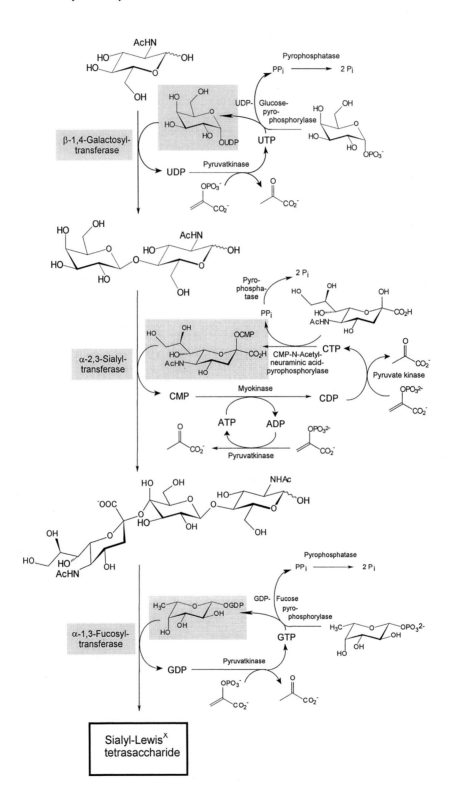

Sialyl-Lewis^x
tetrasaccharide

Normally, one particular glycosyltransferase catalyzes the formation of only a single type of glycosidic linkage, however, the number of exceptions to this rule is increasing. The availability of glycosyltransferases is rather restricted compared to that of the hydrolyzing glycosidases. Increasing the available quantities of these chemically–useful enzymes has been enhanced by techniques such as cloning and overexpression of the proteins in suitable vectors and furthermore, genetic engineering may be used to manipulate their characteristics and specificities.

Transferases require nucleosidediphospho–activated monosaccharides as their donor substrates. These are expensive and their synthesis is difficult and dependent on costly co-factors such as nucleoside triphosphates. These problems have been elegantly solved by expanding enzyme usage to substrate synthesis and continuous cyclic regeneration of the required cofactors. In this way a multitude of reactions can be combined in a multi–step multi–enzyme process for the total synthesis of oligosaccharides. This strategy has been impressively demonstrated by the synthesis of a biologically important tetrasaccharide, called sialyl–Lewis[X]. In this enzymatic multi–step process, the activated substrates, required for enzymatic sugar transfer were also synthesized enzymatically: UDP–galactose, to be transferred by β–1,4–galactosyltransferase; CMP–neuraminic acid, to be transferred by α–2,3–sialyltransferase; and GDP–fucose, which is finally transferred to the intermediate trisaccharide by α–1,3–fucosyltransferase.[25]

Thus it can be seen that the enzymatic synthesis of oligosaccharides is a flourishing field of research with many options and possibilities, for example combining glycosidases with transferases in a sequential process or the use of enzymatic together with chemical methods in so–called chemoenzymatic syntheses.

References

1. J. Dahmén, T. Frejd, G. Grönberg, T. Lave, G. Magnusson, G. Noori (1983) *Carbohydr. Res.* 116, 303.
2. R.R. Schmidt, B. Wegmann, K.-H. Jung (1991) *Liebigs Ann. Chem.* 121.
3. O. W. Lockhoff (1980) Dissertation, University of Hamburg.
4. K.-H. Jung, M. Hoch, R. R. Schmidt (1989) *Liebigs Ann. Chem.* 1099.
5. G. Grundler, R. R. Schmidt (1984) *Liebigs Ann. Chem.* 1826.
6. S. Kötter, U. Krallmann-Wenzel, S. Ehlers, Th. K. Lindhorst (1998) *J. Chem. Soc., Perkin Trans. 1*, 2193.
7. R. Rodebaugh, J. S. Debenham, B. Fraser-Reid, J. P. Snyder (1999) *J. Org. Chem.* 65, 1758.
8. V. Wittmann, D. Lennartz (2002) *Eur. J. Org. Chem.* 1363.
9. T. Wrodnigg, B. Eder (2001) *Topics Curr. Chem.* 215, 115.
10. H. Paulsen, C. Kolár (1981) *Chem. Ber.* 114, 306.
11. M. A. E. Shaban, R. W. Jeanloz (1976) *Carbohydr. Res.* 61, 181.
12. F. Barresi, O. Hindsgaul (1994) *Can. J. Chem.* 72, 1447.
13. Y. Ito, T. Ogawa (1997) *J. Am. Chem. Soc.* 119, 5562.
14. D. Crich, S. Sun (1998), *Tetrahedron* 54, 8321.
15. L. Yan, D. Kahne (1996) *J. Am. Chem. Soc.* 118, 9239.
16. J. Thiem, H. Karl, J. Schwentner (1978) *Synthesis* 693.
17. L. Laupichler, H. Sajus, J. Thiem (1992) *Synthesis* 1133.

18.   K. C. Nicolaou, T. Ladduwahetty, J. L. Randall, A. Chucholowski (1986)
       J. Am. Chem. Soc. 108, 2466.
19.   P. M. St. Hilaire, M. Meldal (2000) *Angew. Chem. Int. Ed.* 39, 1162.
20.   T. Bielfeldt, S. Peters, M. Meldal, K. Bock, H. Paulsen (1992)
       Angew. Chem. Int. Ed. Engl. 31, 857.
21.   H. Paulsen (1982) *Angew. Chem. Int. Ed. Engl.* 21, 155.
22.   G. Hummel, R. R. Schmidt (1997) *Tetrahedron Lett.* 38, 1173.
23.   D. R. Mootoo, P. Konradsson, U. Udodong, B. Fraser-Reid (1988)
       J. Am. Chem. Soc. 110, 5583.
24.   S. C. T. Svensson, J. Thiem (1990) *Carbohydr. Res.* 200, 391.
25.   Y. Ichikawa, J. L. C. Liu, G. J. Shen, C. H. Wong (1991)
       J. Am. Chem. Soc. 113, 6300.

5 Important modifications and functionalizations of the sugar ring

A wide variety of reactions can be effected at carbon atoms other than the anomeric carbon and derivatives prepared from these are often loosely described as 'modified sugars'. They are constituents of many substances with biological activity and pharmacological relevance. Such unusual monosaccharides can be seen in the structures of the enediyne antibiotic[1] calicheamicin γ_1^I and the anticancer drug aclarubicin,[2] for example.

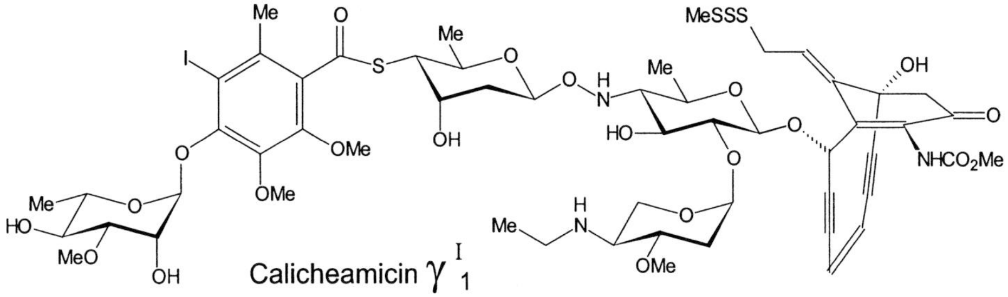

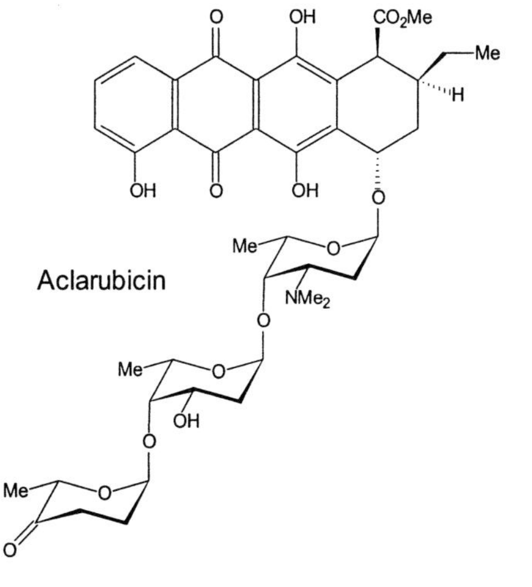

Calicheamicin γ_1^I

Aclarubicin

A number of useful methods to modify and to functionalize carbohydrates are well established and can be considered as essential to carbohydrate chemistry. Typical reaction sequences lead to deoxy, or amino sugars, for example. As well as for the synthesis of complex oligosaccharides in glycosylated natural products such as those depicted above, modified carbohydrates often are designed as enzyme inhibitors of various types for the purpose of investigation of biosynthetic pathways. Moreover, saccharides are frequently derivatized to provide glycomimetics and neoglycoconjugates of various kinds or to produce enantiomerically pure non–carbohydrate molecules from the carbohydrate chiral pool.

5.1   Activation–substitution reactions at the carbohydrate ring

A basic first step in the modification of carbohydrate ring positions is the activation of the sugar hydroxyls as leaving groups. This is mostly accomplished by their conversion into sulfonates, employing the respective sulfonic acid chlorides or anhydrides in pyridine. The most widely used sulfonic esters are mesylates (methanesulfonates), tosylates (*p*–toluenesulfonates), and triflates (trifluoromethanesulfonates). They can be ranked with re-gard to their potency as leaving groups as shown in Table 5–1. Because they can be easily removed, sulfonic esters can be displaced by many types of nucleophiles such as amino, halogen, thio or alkyl derivatives, implying cleavage of the carbon–oxygen bond. These reactions occur with great ease at the anomeric center, relatively readily at primary posi-tions, but with significantly less ease at secondary hydroxyl groups. The displacement reac-tion always occurs by a S_N2 mechanism and, consequently, with inversion of the configura-tion at the substitution center. The reaction conditions selected for most displacements are typical S_N2 conditions, using rather powerful nucleophiles in aprotic polar solvents such as DMF and elevated temperatures up to 100°C.

Table 5–1. Leaving groups which are regularly used in substitution reactions with carbohy-drate derivatives.

	Leaving group	Name
Lower reactivity	Cl–	Chlorine
	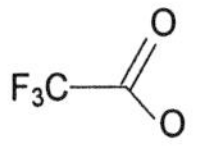 F_3C—C(=O)—O	Trifluoroacetyl
	Br–	Bromine
	I–	Iodine
	H_3C—$S(O)_2$—O—	Mesyl
	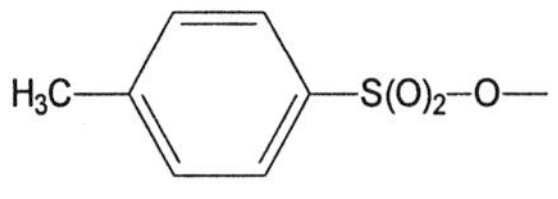 H_3C—C$_6$H$_4$—$S(O)_2$—O—	Tosyl
Higher reactivity	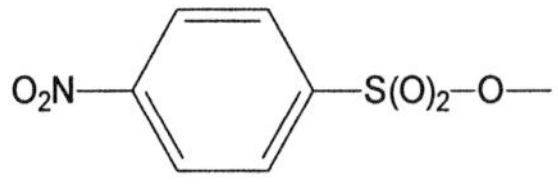 O_2N—C$_6$H$_4$—$S(O)_2$—O—	*p*–Nitrophenylsulfonate
Significantly more reactive than all others	F_3C—$S(O)_2$—O—	Triflyl

As primary hydroxyl groups are more reactive than those at secondary positions, the 6–positions of monosaccharides can easily be selectively tosylated or mesylated for example, using the respective sulfonyl chlorides in pyridine or pyridine–dichloromethane mixtures at approximately 0°C. The resulting 6–sulfonates can eventually be converted to 6–deoxy–6–iodo or 6–azido–6–deoxy derivatives by nucleophilic substitution. The iodo derivatives are precursors for deoxy compounds and the azides can be regarded as masked amines. Tosylates can also be converted into the respective deoxy derivatives in a direct reaction, using lithium aluminum hydride at room temperature.

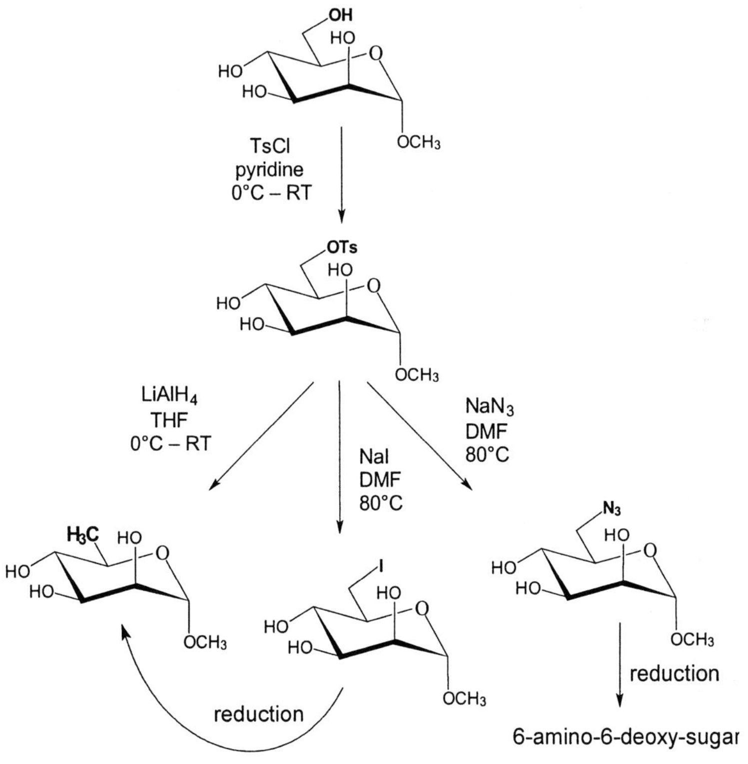

Synthesis of methyl 6–*O*–tosyl–α–D–mannopyranoside.[3,4] A solution of 15.2 g methyl α–D–mannopyranoside (77.9 mmol) in 60 ml of dry pyridine is cooled to 0°C and then a solution of 1.1 equivalents of tosyl chloride in 30 ml of dry pyridine is added dropwise at this temperature. The reaction mixture is allowed to warm to room temperature and is continually monitored by TLC. Once the reaction is complete, the volume is doubled with water and the aqueous phase is extracted three times with ethyl acetate. The combined organic phases are then washed with 5% aqueous HCl solution, followed by water and then dried over MgSO₄. After filtration, the solvent is removed *in vacuo* and the residue purified by flash chromatography. The tosylated saccharides prepared in this way can be converted to the respective iodides or azides by stirring in anhydrous DMF solution with five equivalents of sodium iodide or sodium azide respectively, at 80°C until the reaction is complete. For work–up, almost all the DMF is removed under high vacuum and the residue is extracted between water and ethyl acetate and the usual procedure followed.

Secondary hydroxyl groups can also be converted into sulfonyl esters. However, the re-activity of secondary positions in nucleophilic substitution reactions is often low, especially in the case of axially–oriented hydroxyl groups, triflates (trifluoromethanesulfonates) are therefore used as leaving groups to enhance the reaction.

To synthesize the triflate derivatives of sugars, the starting material is dissolved in dry dichloromethane and a small amount of pyridine and is reacted with triflic anhydride (tri-fluoromethanesulfonic anhydride) at a low temperature (-10°C). Carbohydrate triflates are apparently quite stable up to certain temperatures beyond which they degrade quickly. They are often used as raw materials for subsequent substitution reactions, but can normally also be purified on silica gel to yield the products some of which are crystalline.

General procedure for the synthesis of triflate esters from alcohols and triflic an-hydride. In a 100–ml, two–neck, round–bottom flask, equipped with two dropping funnels pyridine (0.43 ml, 5.5 mmol) and 20 ml dichloromethane are kept under an inert atmosphere (N_2 or Ar). One funnel is filled with a solution of triflic anhydride (0.86 ml, 5.11 mmol) in 10 ml of dichloromethane and the other with the sugar (2.55 mmol) dissolved in 10 ml of dichloromethane. The flask is cooled to -10°C and the triflic anhydride solution is added dropwise. A thick white precipitate is formed during the addition. After the addition is com-plete, the suspension is allowed to stir for 10 minutes before the sugar solution is added dropwise. Stirring at –10°C is continued for approx. 1.5 h until the conversion of the sugar alcohol is complete (monitored by TLC). The reaction mixture is then poured onto 50 ml of ice water, the layers are separated and the aqueous layer is extracted with two 50 ml–portions of dichloromethane. The combined extracts are dried ($MgSO_4$), filtered and the sol-vent removed *in vacuo*. In most cases the product can be used in the next step without fur-ther purification.

For the synthesis of 1,2,3–tri–*O*–benzoyl–4,6–dideoxy–4–iodo–α–L–glucopyranose, a precursor of 4–deoxy–L–fucose, a triflate intermediate prepared from 1,2,3–tri–*O*–benzoyl–6–deoxy–α–L–galactopyranose is the only suitable derivative for the substitution reaction with iodide.

Synthesis of 1,2,3–tri–*O*–benzoyl–4,6–dideoxy–4–iodo–α–L–glucopyranose.[5]

The corresponding fucose derivative with the 4–OH group unprotected (4.66 g, 9.78 mmol) is dissolved in 50 ml of dry dichloromethane and 6.5 ml of dry pyridine. With the exclusion of moisture a solution of 4.0 ml (23.77 mmol) of triflic anhydride in 7 ml of dichloromethane is added dropwise at -20°C. The reaction mixture is then stirred at room temperature until the alcohol is completely converted to the triflate (monitored by TLC). Water (50 ml) is then added and the aqueous phase is extracted three times with dichloromethane. The combined organic phases are co–evaporated with toluene *in vacuo* at low temperature. The resulting residue is immediately dissolved in 50 ml of dry DMF, 4.0 g (26.7 mmol) of sodium iodide are added and the reaction mixture is stirred overnight at room temperature. The solution is then reduced to half the volume *in vacuo*, water is added and the aqueous phase is extracted four times with dichloromethane. The combined organic phases are washed with water twice and then co–evaporated with toluene to dryness. After chromatographic pre–purification the product can be readily crystallized from dichloromethane to obtain the inverted 4–deoxy–4–iodo compound as colorless crystals in 87% yield.

Triflyloxy groups are the most easily displaced groups and are widely used for problematic syntheses. However, triflate substitutions are sometimes accompanied by elimination reactions leading to unsaturated carbohydrate derivatives as side products (*vide supra*). On the other hand they also offer the advantage of direct reduction to the respective deoxy sugars, using hydride reagents.

The difference in reactivity between primary and secondary sulfonates can be utilized in selective displacement reactions.

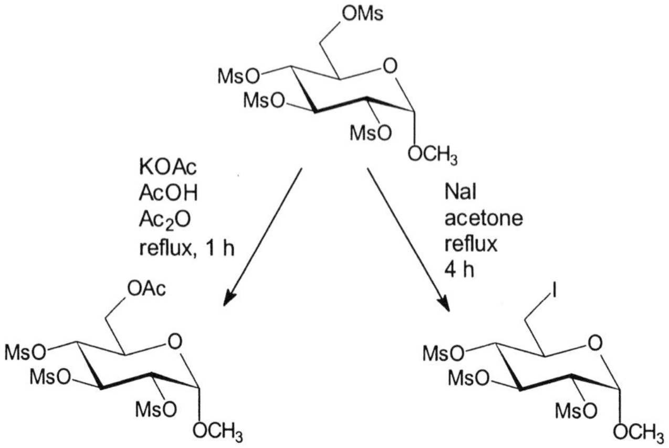

Intramolecular substitution reactions of tosylated carbohydrates can be readily used for the preparation of sugar oxolanes and oxiranes (epoxides) which are important building blocks in carbohydrate synthesis.

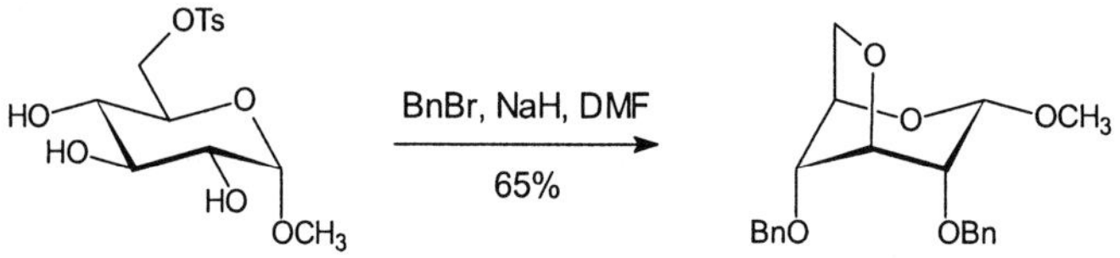

3,6-anhydro sugar
(an oxolane)

Synthesis of methyl 2,3–anhydro–4,6–O–benzylidene–α–D–mannopyranoside.[6]

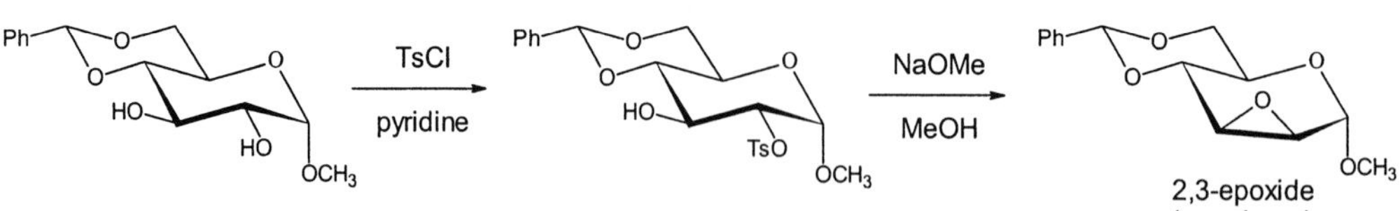

2,3-epoxide
(an oxirane)

A hot suspension of 1.5 g (3.4 mmol) of methyl 4,6–O–benzyliden–2–O–benzyl–α–D–glucopyranoside in 25 ml of dry methanol is treated with a solution of 0.9 g (20 mmol) sodium methanolate in 5 ml of dry methanol. The reaction mixture is then heated under reflux for approx. 1h. Upon cooling to room temperature a solid begins to precipiate. The mixture is poured onto ice water (200 ml), the precipitate is filtered off, washed with water, and dried in a dessicator over phosphorous pentoxide. The target epoxide (0.8 g, 89%) is obtained as a white solid.

Many other important substitution reactions start from acetylated glycosyl bromides. These are obtained by stirring the respective peracetates in HBr–acetic acid solution. The bromine atom in glycosyl bromides is already activated towards nucleophilic substitution because of its anomeric position. Substitution with azide anions in a solvents such as DMF or acetonitrile results in the corresponding glycosyl azide in high yield; the azide being a precursor for glycosyl amines. The latter are, like aminals, not very stable and have to be prepared freshly before further conversion to glycopeptides, for example. Reaction of glycosyl bromides with thiocyanate anions leads to the glycosyl isothiocyanates, which are important precursor molecules for *N*–glycosyl thiourea derivatives. The formation of glycosyl isothiocyanates is, however, frequently complicated by the concomitant formation of the isosteric glycosyl thiocyanates due to the ambident character of the SCN⁻ anion. As the thiocyanates rearrange to isothiocyanates under thermodynamically controlled conditions, selective formation of glycosyl isothiocyanates can be effected by reacting the glycosyl bromides and an excess of KSCN in the melt for 5 minutes. Glycosyl thiols are obtained from the corresponding bromides in the reaction with thiourea, and reduction of the resulting isothiouronium bromides.

Synthesis of 2,3,4,6–O–acetyl–α–D–glucopyranosyl bromide.[7] To 20 g (51 mmol) of penta–O–acetyl–β–D–glucopyranose, 50 ml of 33% HBr–HOAc solution and 10 ml of acetic anhydride are added at 0°C and then the reaction mixture is stirred at room temperature for 2 h. 200 ml of dichloromethane are added and the solution is poured onto 300 ml of ice–water. The phases are separated and the aqueous phase is twice extracted with dichloromethane. The organic phases are combined and carefully neutralized with satd. sodium hydrogen carbonate solution, washed with water and dried over $MgSO_4$. After filtration and evaporation of the solvent at low temperature, the residue is dissolved in a minimum amount of diethyl ether and crystallized by addition of little petroleum ether. 17.3 g (83%) of the product are obtained in form of colorless crystals.

5.2   Deoxyhalo derivatives

Deoxyhalo sugars are important precursors for deoxygenated carbohydrates, i.e. they can be easily obtained by a suitable displacement reaction from carbohydrate sulfonates and this is readily achieved in the case of iodide and bromide. However, from the usual order of reactivity of halide ions in S_N2 displacements in solution, iodide > bromide > chloride > fluoride, it can be seen that the latter is a rather weak nucleophile. On the other hand, the preparation of deoxyfluoro sugars, in which one of the ring–hydroxyl groups (except the anomeric hydroxyl) has been substituted by fluorine, is of particular interest as these sugars are of great biological interest. They serve as probes for the binding sites of anti–saccharide antibodies and for the active sites of enzymes and are therefore used to investigate their catalytic mechanisms and specificities. Deoxyfluoro sugars are especially useful derivatives in this context because, firstly, the substitution of a sugar hydroxyl by a fluorine is sterically conservative since fluorine is smaller than the original hydroxyl. And secondly, the fluorine cannot donate a hydrogen bond but it can act as a hydrogen bond acceptor. Thus, testing a series of deoxyfluoro–modified carbohydrate derivatives for their binding capacity to a receptor or an enzyme will provide useful information about the hydrogen–bonding interactions necessary for molecular recognition. This information can be further supplemented by probing an analogous series of deoxygenated derivatives, in which the deoxy functions do not participate in hydrogen bonding. Moreover, binding studies with fluorinated carbohydrate derivatives can be substantiated by ^{19}F NMR spectroscopic studies.

Fluoride displacements at secondary positions are the most difficult to achieve on account of the weakly nucleophilic character of this ion. Sources for fluoride are usually caesium fluoride or tetrabutylammonium fluoride, but bis(tetrabutylammonium) difluoride $(Bu_4N)_2F_2$ and tris(dimethylamino) sulfur (trimethylsilyl)difluoride $[(Me_2N)_3S(Me_3SiF_2)]$

(TASF) can also be used successfully. To facilitate nucleophilic displacement the strongly electron–withdrawing trifluoromethylsulfonyl group must be used as the leaving group in this reaction, which is unfortunately, often accompanied by undesired elimination reaction favored by the somewhat basic character of the fluorine anion.

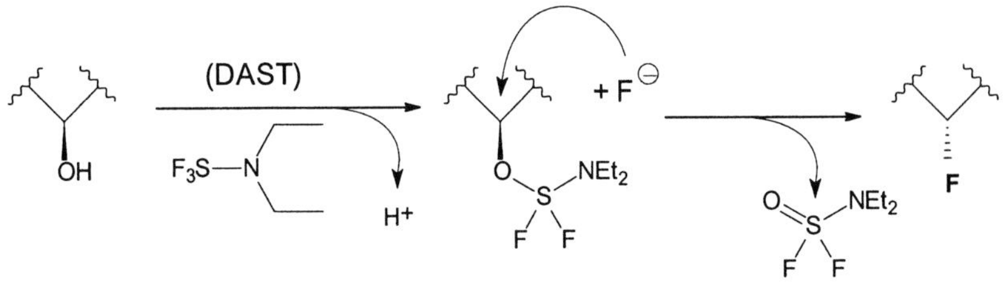

Alternatively, alcohol groups can be directly replaced by fluorine atoms using DAST (di–ethylaminosulfur trifluoride) as the fluorinating reagent. It can in principle be used for the direct fluorination of all hydroxyls of the sugar ring, including the anomeric hydroxyl, just by stirring at room temperature. Mostly, fluorination with DAST occurs with inversion of the configuration at the center of the nucleophilic substitution. This finding is in agreement with an activation–displacement mechanism which has been suggested for the reaction of hydroxyl groups with DAST, which is however, not the only pathway the reaction may take. Using DAST, elimination in the fluorination reaction can be substantially diminished.[8]

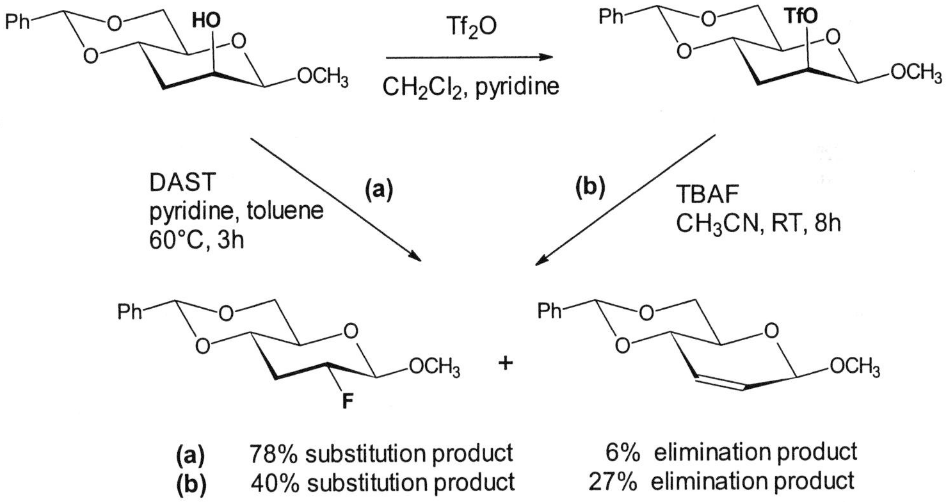

(a)   78% substitution product   6% elimination product
(b)   40% substitution product   27% elimination product

DAST can also be used for regioselective fluorination. Methyl α–D–glucopyranoside gives the 6–deoxy–6–fluoro derivative after a 15–minute treatment with DAST, but pro–longed treatment furnishes the 4,6–dideoxy–4,6–difluoro–galactoside. The major product from prolonged reaction of DAST with methyl β–D–glucopyranoside leads to the 3,6–difluorinated allopyranoside.

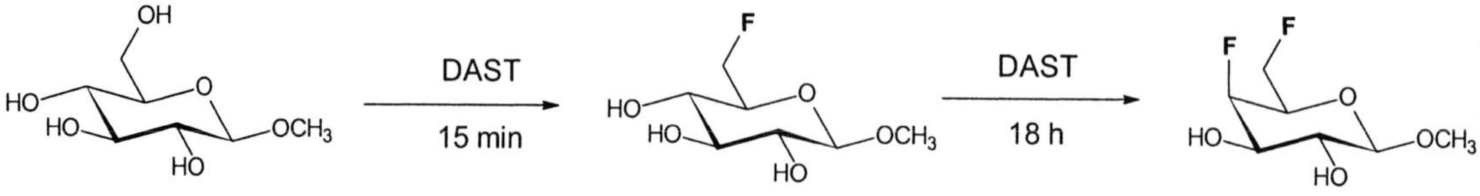

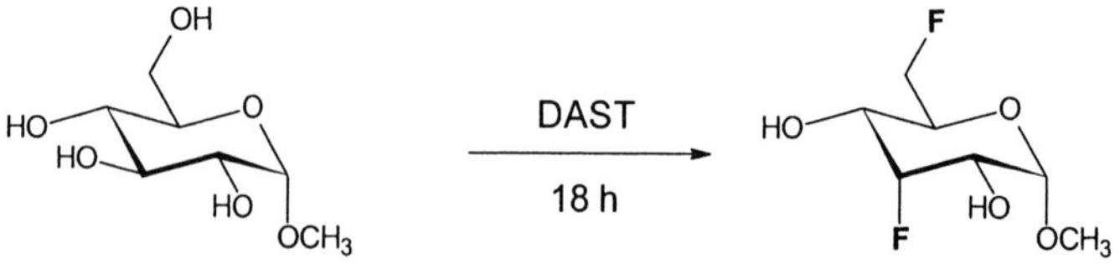

Fluorination of the 2–hydroxyl with DAST sometimes leads to undesired results due to migration of the anomeric substituent to the 2–postion and fluorination occurring at the anomeric center, resulting in the formation of glycosyl fluorides. This side reaction has been described for methyl glycosides, glycosyl acetates, glycosyl azides as well as for thio glycosides.

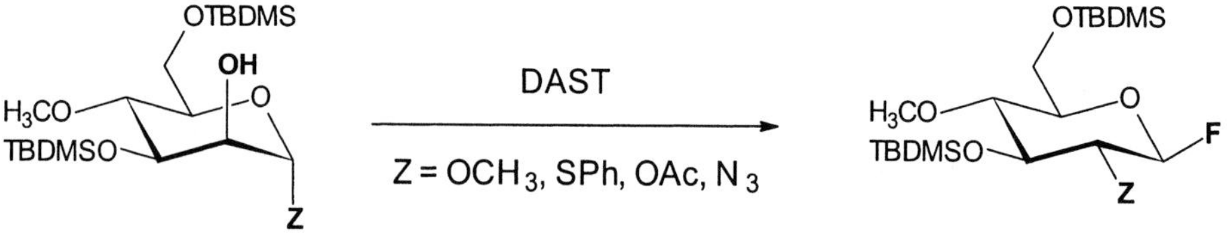

The fact that the C–F bond is not reduced by Bu₃SnH can be utilized for the synthesis of deoxygenated glycosyl fluorides, which can then be used as glycosyl donors. Thus, a 2,6–dibromo–2,6–dideoxy glycosyl bromide can be converted to the glycosyl fluoride by halogen exchange and can then be reduced to the corresponding 2,6–dideoxy glycosyl fluoride.[9]

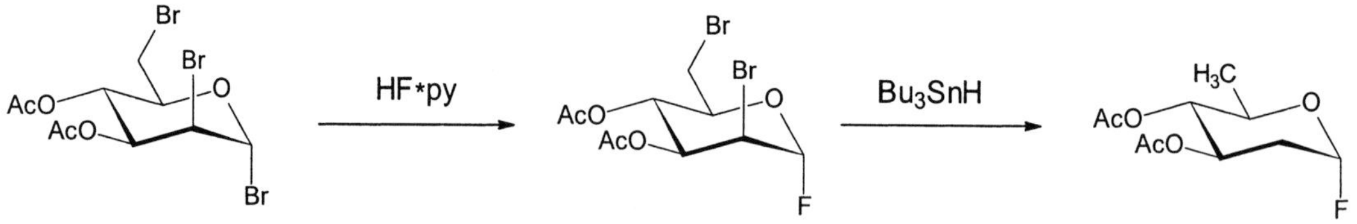

Other halogen atoms can also be introduced into the carbohydrate ring by direct displacement of hydroxyl groups, instead of using an activation–substitution protocol. 6–Chloro–6–deoxy glycosides can be obtained upon treatment of the unprotected glycoside with mesyl chloride in DMF, via the formation of an intermediary imminium salt.

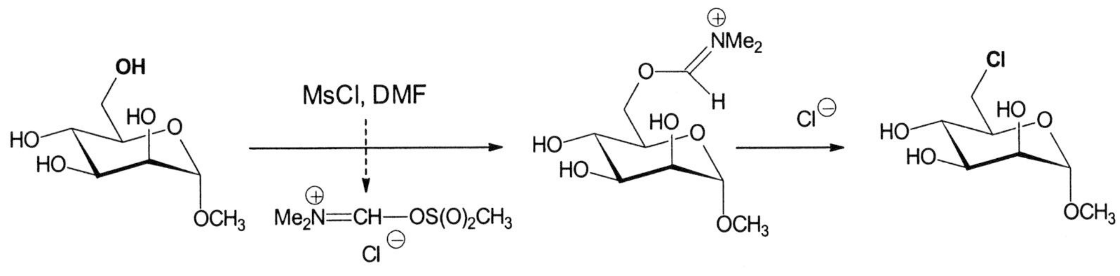

General procedure for the synthesis of 6–deoxy–6–halo–glycosides. Methylsulfonyl chloride (10 eq.) is added dropwise to a stirred solution of the glycoside (1 eq.) in anhydrous DMF (0.1 g/ml) and the reaction mixture is maintained for 16 h at 65°C. It is then concentrated to a syrup which is dissolved in MeOH and treated with sodium methoxide to destroy O–formate esters. The solution is concentrated and the product purified on silica gel and crystallized.

The application of an Appel reaction[10] using PPh$_3$ together with CCl$_4$ or CBr$_4$ on unprotected glycosides, is another method of selective 6–halogenation. Thus, the treatment of unprotected glycosides with NBS or NIS in the presence of triphenylphosphine in DMF yields 6–deoxy–6–halo derivatives. A modification of this method has been introduced by P. Garegg in which the unprotected glycoside with PPh$_3$, imidazole and iodine in toluene is refluxed for approximately 6 hours to obtain the regioselected 6–iodinated monosaccharide.[11]

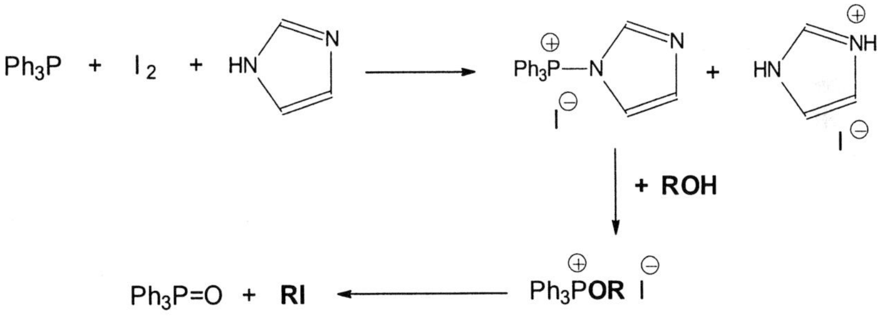

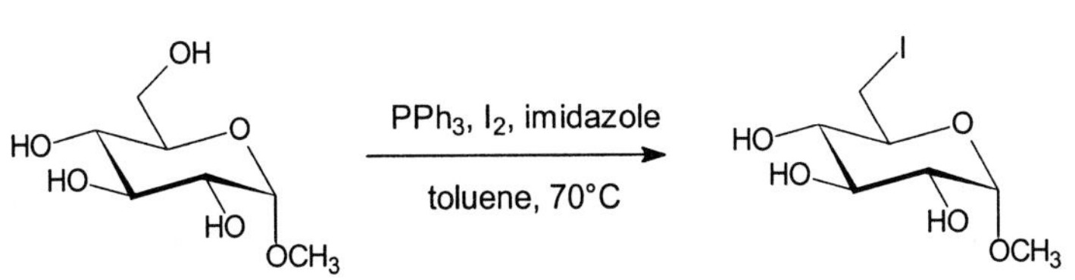

General mechanism for the conversion of alcohols into alkyl iodides using PPh3, imidazole, and iodine. This procedure can be applied for the regioselective iodination of the 6–position of unprotected glycosides.

5.3   Deoxygenation of carbohydrates

Replacement of one or more hydroxyl groups of a carbohydrate leads to derivatives, called deoxy sugars. Like deoxyfluoro sugars, deoxygenated carbohydrates vary in their ability to act as ligands for lectins or substrates for enzymes, and because of this can be used for the elucidation of the binding modes and mechanisms. Deoxygenations can be achieved by many different procedures such as

(i)   the reduction of deoxyhalo sugars;
(ii)   opening of epoxides;
(iii) the reduction of esters;
(iv)   the radical reduction of sulfur–containing derivatives;
(v)   by addition to unsaturated derivatives such as glycals or
(vi)   reduction of unsaturated derivatives to vicinal dideoxy sugars.

Deoxyiodo and deoxybromo sugars can be reduced to the deoxy analogs by employing radical reduction with tributylstannane, for example, or by Pd–catalyzed hydrogenation or using Raney–nickel instead of palladium. Deoxychloro sugars are easily reduced with lithium aluminum hydride. Synthesis of multiply deoxygenated sugars may practically combine the Garreg reduction of a vicinal diol group and a Hanessian–Hullar reaction, in which ring cleavage of carbohydrate benzylidene acetals with *N*–bromosuccinimide leads to benzoylated deoxybromo sugars, which can then be readily reduced.

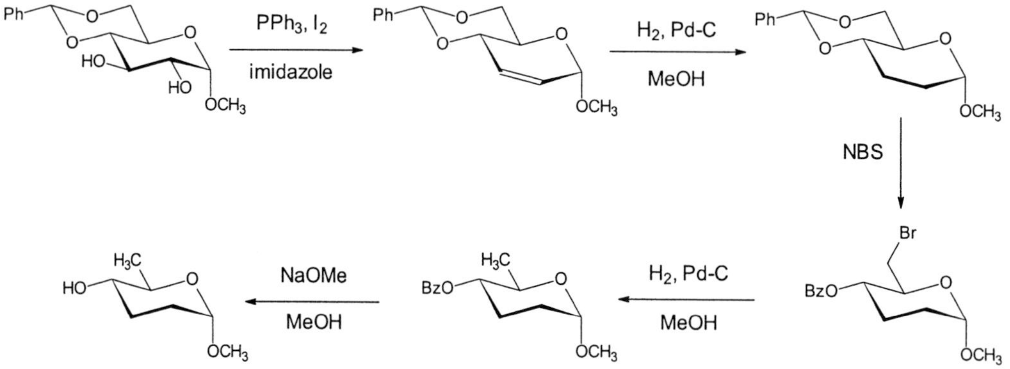

Deoxygenated derivatives can also be obtained from epoxides. Epoxide ring–cleavage of 2,3–anhydro–hexopyranosides with nucleophiles leads mainly to products having C–2 and C–3 substituents in the *trans*–diaxial orientation. This phenomenon is often referred to as the Fürst–Plattner rule. Using lithium aluminum hydride for epoxide ring–opening, deoxy derivatives are obtained. Often one of four possible stereoisomeric products predominates. With methyl 2,3–anhydro–4,6–*O*–benzylidene–α–D–allopyranoside as the starting material only the *ribo*–configured 2–deoxy derivative having an axial substituent at C–3 is produced, whereas with the *manno*–configured analog the 3–deoxy sugar is obtained.

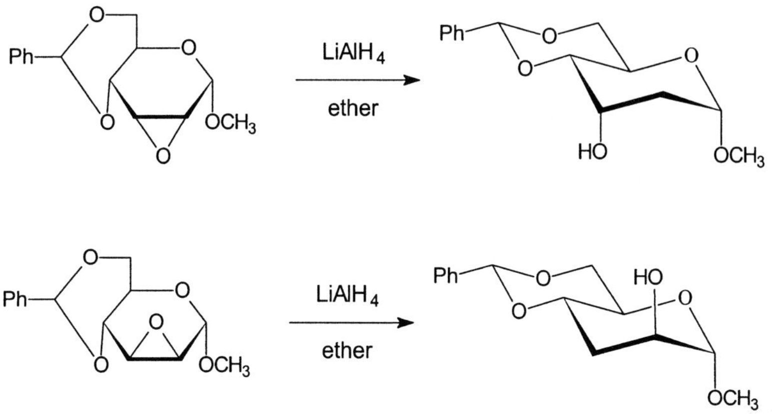

Synthesis of methyl 4,6–O–benzylidene–3–deoxy–α–D–arabinohexopyranoside.[12] A solution of 0.5 g (1.9 mmol) methyl 2,3–anhydro–4,6–O–benzylidene–α–D–mannopyrano-side in 20 ml of dry THF is treated with 0.30 g (8 mmol) of lithium aluminum hydride and stirred under reflux for 4 h. Then the reaction mixture is cooled to 0°C and 1 ml of water is carefully added, followed by addition of 1 ml of 15% aqueous NaOH solution. The mixture is then stirred at room temperature for 1 h, filtered over celite and evaporated *in vacuo*. The colorless syrup obtained is purified by flash chromatography and the purified product starts to crystallize after a few days. The yield is 82% (0.42 g).

An especially suitable method for the deoxygenation of secondary alcohols is the Barton–McCombie reaction,[13] in which thiocarbonyl derivatives of sugar hydroxyl groups are reduced to the deoxy function in a radical reduction using tributyl tin hydride. This type of reaction is useful with a variety of differently functionalized hydroxyl groups such as xanthates, thiocarbonylimidazolides, or phenylthionocarbonates.

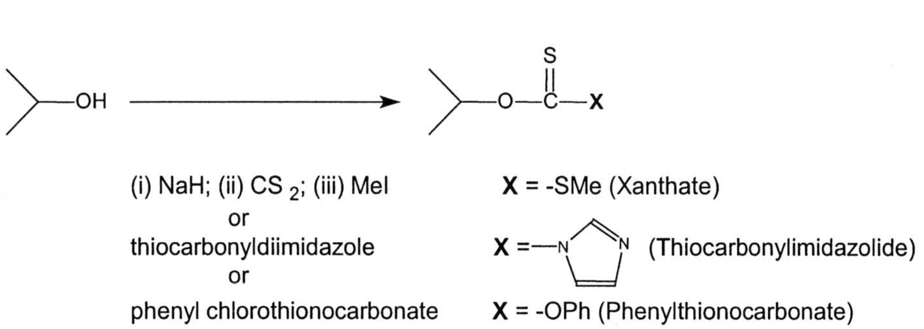

(i) NaH; (ii) CS$_2$; (iii) MeI   X = -SMe (Xanthate)
            or
thiocarbonyldiimidazole   X =—N⁀N (Thiocarbonylimidazolide)
            or
phenyl chlorothionocarbonate   X = -OPh (Phenylthionocarbonate)

The mechanism of this radical chain reaction proceeds as follows. The thiocarbonyl derivative affords an intermediate radical, following attack by the tributyl tin radical, with the formation of a tin–sulfur bond. Then fragmentation into the desired carbon radical and a thiocarbonylic derivative takes place. The carbon radical is finally reduced by hydrogen atom transfer to the desired deoxy compound with regeneration of the tributyl tin radical.

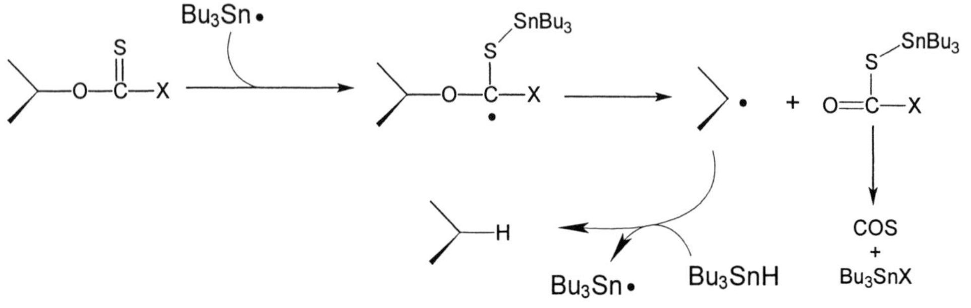

A practical example of radical deoxygenation is the synthesis of 3–deoxy–L–fucose starting with a fucoside, that has only the 3–position unprotected.

Synthesis of methyl 2,4–di–O–benzoyl–6–deoxy–3–O–(1–imidazolylthiocarbonyl)– α–L–galactopyranoside (step (i)).[14]

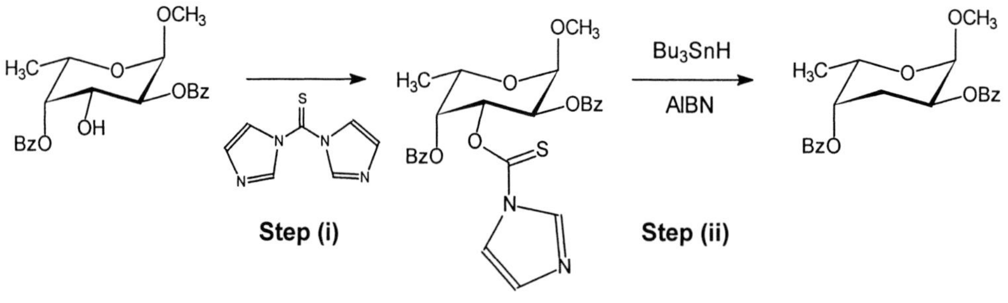

A solution of 1 g (2.58 mmol) of methyl 2,4–di–O–benzoyl–6–deoxy–α–L–galactopyranoside in 100 ml of dry dichloromethane is treated with 930 mg (5.2 mmol) of 1,1–thiocarbonyldiimidazole and heated under reflux. The product formed cannot be distinguished from the starting material on TLC. After 8 h reaction time the organic solvent is removed *in vacuo* and the residue is purified on silica gel to yield 1.2 g (97%) of the desired product as colorless needles.

Synthesis of methyl 2,4–di–O–benzoyl–3,6–dideoxy–α–L–lyxo–hexopyranoside (step (ii)).[14] A solution of 820 mg (1.65 mmol) of methyl 2,4–di–O–benzoyl–6–deoxy–3–O–(1–imidazolylthiocarbonyl)–α–L–galactopyranoside in 15 ml of anhydrous toluene is added dropwise to a refluxing mixture of 0.9 ml (3.4 mmol) of tri–n–butylstannane, and a catalytic amount of AIBN in 40 ml of dry toluene. The reaction mixture is stirred under reflux for 4 h. The organic solvents are then evaporated off and the residue is diluted with n–hexane and extracted five times with acetonitrile, to remove alkyl stannan side products with the hexane phase. The combined acetonitrile phases are concentrated and the residue purified by flash chromatography to yield 420 mg (69%) of the desired deoxy derivative as colorless syrup.

Alternative reagents to tributyl tin hydride are the less toxic diphenylsilane or tris(trimethylsilyl)silane, either of which can be used in the radical reduction reaction.

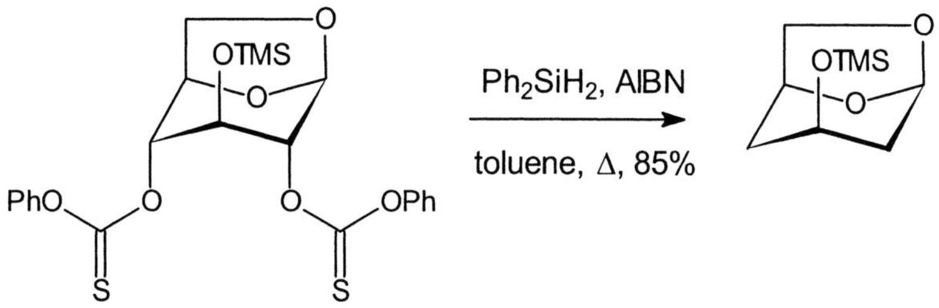

5.4   Unsaturated saccharides – Glycals

A wide range of unsaturated monosaccharide derivatives are known and serve as versatile building blocks in carbohydrate chemistry and also as starting materials for the preparation of non–carbohydrate products. Naming unsaturated carbohydrates according to IUPAC nomenclature is somewhat annoying and, therefore, the simpler trivial names are often used. Of particular synthetic significance are monosaccharides with a double bond between C–1 and C–2, the 1,5–anhydro–2–deoxy–hex–1–enitols, also called glycals.

Glycals were given the suffix 'al' by Emil Fischer who was misled by a positive Fuchsin–SO_2–test on some crude material he had prepared, thinking that he had synthesized an aldehyde. The culpable aldehyde, however, is an open–chain derivative, which is produced upon prolonged hydrolysis of the acetylated glucal.

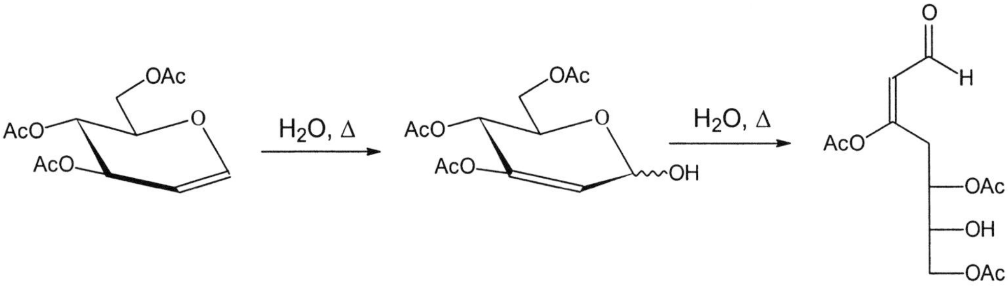

Acetylated glycals are traditionally prepared by the action of zinc in acetic acid on acetylated glycosyl halides. These conditions are typical of soluble metal reductions, and the elimination may occur following reduction of C–1 carbocations by the metal. Mild base–catalyzed deacetylation yields the unprotected glycal.

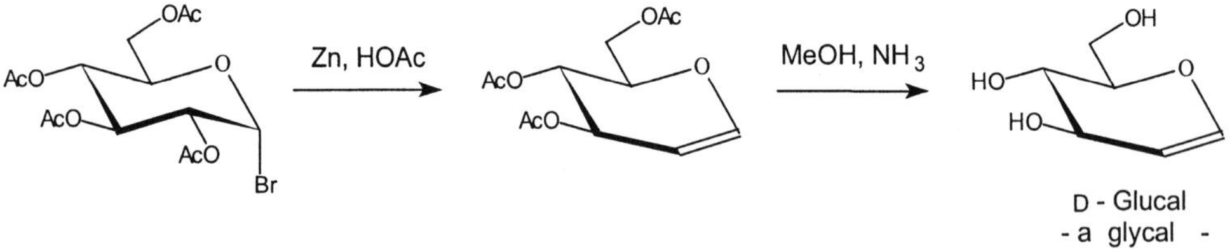

D - Glucal
- a glycal -

The zinc–based method has its limitations in that it can be somewhat cumbersome to apply, and the conditions are too aggressive for use in making furanoid glycals, for example. Many alternatives have been described for the preparation of glycals, including reaction under aprotic conditions using Zn–Ag–graphite in THF at -20°C or zinc and 1–methylimidazole in ethyl acetate under reflux. The latter method has been shown to work well for the synthesis of the disaccharidic cellobial.

Synthesis of 1,5–anhydro–3,6–di–O–acetyl–4–O–(2,3,4,6–tetra–O–acetyl–β–D–glucopyranosyl)–2–deoxy–D–arabino–hex–1–enitol (hexa–O–acetyl–cellobial).[15] To a refluxing suspension of 5.2 g (79.2 mmol) of zinc powder and 1.1 ml (13.2 mmol) of 1–methylimidazole in 70 ml of ethyl acetate, a solution of 9 g (13.2 mmol) of peracetylated cellobiosyl bromide in 15 ml of ethyl acetate is added dropwise over 1 h. The reaction mixture is stirred under reflux until the reaction is complete (monitored by TLC), then it is allowed to cool to room temperature, filtered over celite, washed with 5% aqueous HCl and saturated aqueous NaHCO$_3$ solution and dried over MgSO$_4$. After filtration, the solvent is removed and the crude product purified by flash chromatography to yield 4.1 g (55%) of the disaccharide glycal as colorless crystals.

Glycals with substituents at C–2 are called hydroxyglycals. They have classically been obtained by the elimination of HBr from acetylated glycosyl bromides using diethylamine. In these cases and also when other secondary amines are employed as bases, the elimination reaction is accompanied by nucleophilic substitution leading to *N*–glycosides. The use of DBU is a good alternative, which leads to hydroxyglycals in excellent yields.

Synthesis of 1,5–anhydro–2,3,4–tri–O–acetyl–6–deoxy–2–hydroxy–L–talo–hex–1–enitol (tri–O–acetyl–2–hydroxy–L–fucal).[16]

2,3,4-Tri- O-acetyl-2-hydroxyfucal
- a hydroxyglycal -

To a solution of 650 mg (1.8 mmol) of 2,3,4–tri–O–acetyl–L–fucopyranosyl bromide in 10 ml of dry dichloromethane 2 ml of DBU are added and the reaction mixture is stirred at room temperature for 2 h. It is then poured onto ice–cold 2 N aqueous sulfuric acid solution and extracted with three 30–ml portions of dichloromethane. The combined organic phases are washed with water and dried over MgSO$_4$. After filtration the solvent is evaporated and the solid residue is recrystallized from methanol to yield 450 mg (90%) of the desired hydroxyfucal as colorless needles.

Reactions of (hydroxy)glycals

Many sophisticated synthetic pathways originate from glycals as well as hydroxyglycals often leading to non–carbohydrate natural products. This is due to the special reactivity of these compounds. Whereas carbohydrates with double bonds at positions other than between C–1 and C–2, exhibit normal alkene chemistry, glycals are vinyl ethers and therefore undergo a number of highly selective addition reactions due to the strongly polarized double bond. In general, dissociable species of the kind A$^+$B$^-$ add across the double bond of the enol–ether type–glycals (1,2–dideoxy–hex–1–enopyranoses) leading to 2–deoxy–glycopyranose derivatives and 2–deoxy–glycosides, respectively, in the case of addition of alcohols.

In the addition of halogens to glycals, of the four possible isomeric 1,2–addition products of the reaction, the α–anomers (carrying the halogen atom at C–1 in the axial position) are generally predominant because they are more stable as a result of the anomeric effect. The 1,2–dideoxy–dihalogen products can be used as glycosyl donors leading to alkyl 2–deoxy–glycosides after reductive dehalogenation at C–2.

Likewise, addition of hydrogen halides across the double bond of glycal derivatives yield the 2–deoxy–α–glycosyl halides again as a result of the anomeric effect. These compounds can also be used as glycosyl donors to form 2–deoxy glycosides.

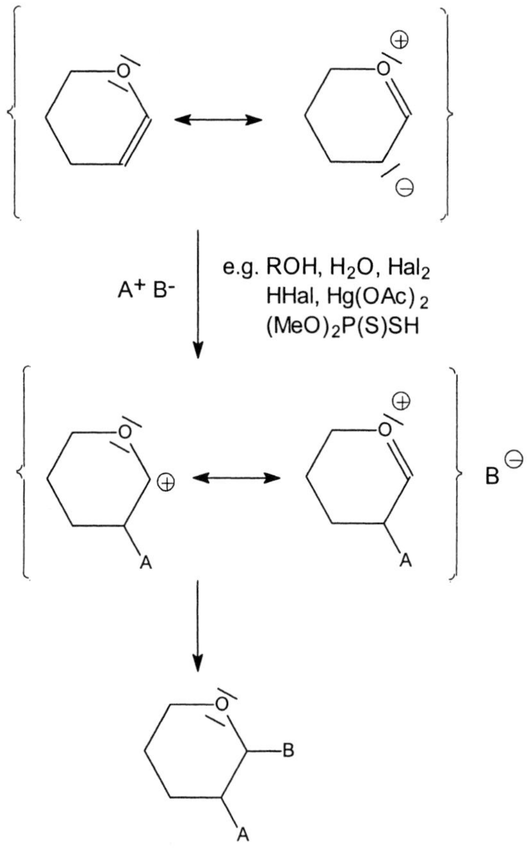

Addition to the double bond can also be effected by the so–called azidonitration reaction introduced by R. Lemieux in 1979.[17] This reaction occurs with sodium azide and cerammonium nitrate resulting in a 2–azido–2–deoxy derivative with an anomeric nitrate group, which can be hydrolyzed or replaced by a halide en route to becoming a glycosyl donor for 2–deoxy–2–amino sugars.

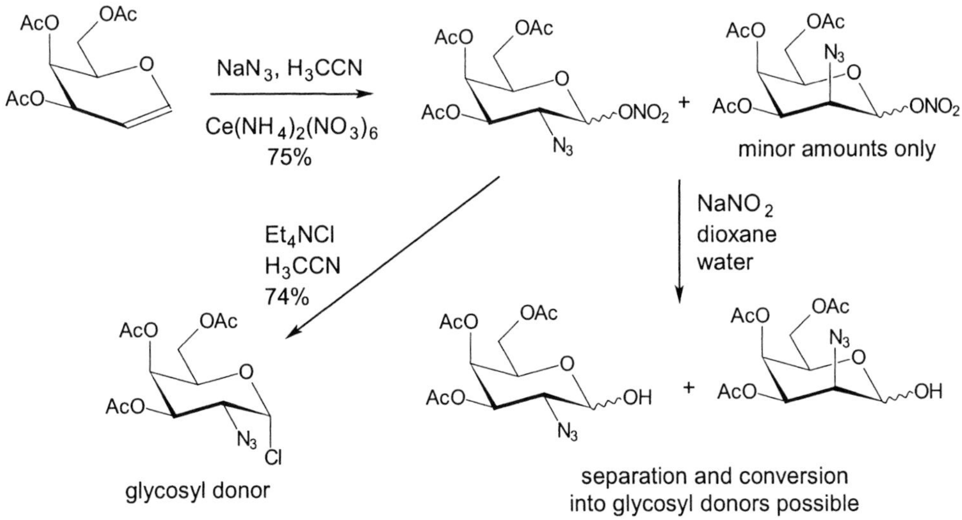

Azidonitration with acetylated galactal leads to the 2–azido–2–deoxy derivatives.

The addition of nucleophiles to glycals in the presence of mineral or Lewis acid (e.g. $BF_3 \cdot EtO_2$) catalysts is generally accompanied or even replaced by concurrent allylic rearrangement involving the elimination of the C–3–acyloxy leaving group as a result the introduction of a nucleophilic group at C–1. This reaction leading to hex–2–enopyranosides has been intensively investigated by Ferrier and is therefore often referred to as the Ferrier rearrangement.[18] A variety of reagents, including alcohols, triethylsilane, pyridinium poly(hydrogen fluoride), nitrogen nucleophiles such as sodium azide, or trimethylsilyl cyanide can be used in this reaction leading to unsaturated O–, N–, S–, or C–glycosides, respectively. α,β–Mixtures are obtained in the Ferrier rearrangement reaction depending on the structure of the starting materials.

Synthesis of Methyl 4,6–di–O–acetyl–2,3–dideoxy–α–D–erythro–hex–2–enopyranoside.[19]

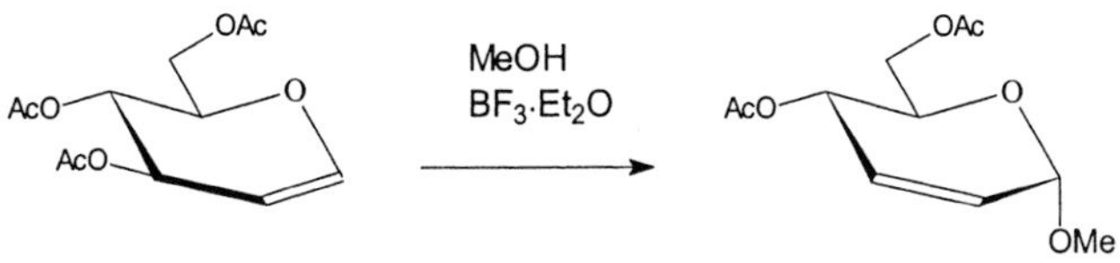

To a solution of 10 g (36.7 mmol) tri–O–acetyl–glucal in 100 ml of dry dichloromethane 35 ml of a mixture consisting of $BF_3 \cdot Et_2O$, methanol, and dichloromethane (1:1:8) is added dropwise with the exclusion of moisture. After 45 minutes the reaction mixture is diluted with 50 ml of dichloromethane and washed with satd. aqueous $NaHCO_3$ solution. The organic phase is dried over $MgSO_4$, filtered and evaporated. After flash chromatography 7.26 g (81%) of the syrupy rearrangement product are obtained.

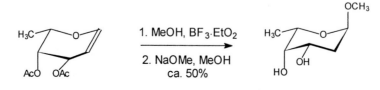

In the Ferrier rearrangement a resonance–stabilized allyloxocarbenium ion is formed under the influence of a Lewis acid catalyst, which can be attacked by a nucleophile in a S$_N$1′–type reaction. This is generally occurs at the more electropositive C–1 with relatively strong nucleophiles (bases) such as OR$^-$ or F$^-$.

The Ferrier rearrangement cannot be applied universally and the outcome of the reaction is strictly dependent on the reagents. For 3,4,5–tri–*O*–acetyl–D–galactal and 3,4–di–*O*–acetyl–L–fucal, which contain the C–3 and C–4–acetoxy functions in a *cis* relationship, the allylic rearrangement reaction is less favored and the 1,2–addition of alcohols is strongly predominant.

The rearranged 2,3–unsaturated pyranosides normally have oxygen–bonded groups at the allylic position and therefore are often involved in allylic displacements and in (retro) rearrangement reactions. Thus, hex–2–enopyranosides, the products of the Ferrier rearrangement, rearrange upon treatment with lithium aluminum hydride to the respective 1,5–anhydro–2,3–dideoxy–hex–1–enitol in many cases. Furthermore, rearrangement reactions with unsaturated monosaccharide derivatives give rise to a variety of reaction pathways and unexpected products. Esters of 2–hydroxyglycals, for example, are suitable sources of several pyranoid enone derivatives, which can be obtained with *m*-chloroperbenzoic acid as shown.

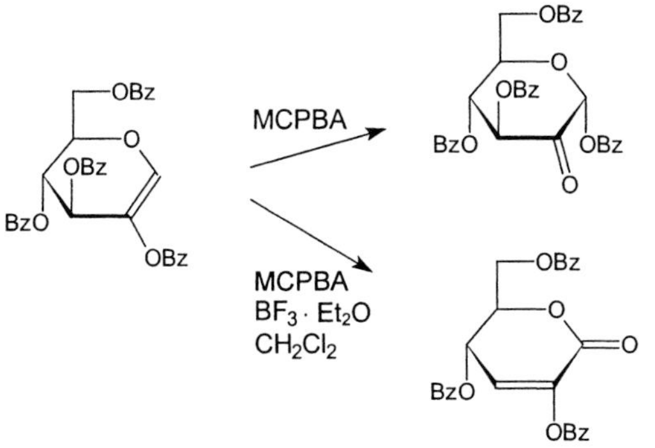

Oxidation of the glycal double bond leads to 1,2–epoxides. Such derivatives were introduced at the beginning of the century and became known as 'Brigl's' anhydrides. They have been subsequently used as glycosyl donors and Schuerch and co–workers demonstrated their usefulness in polymerization reactions. A facile synthesis of 1,2–anhydro sugars from glycals involves the use of 3,3–dimethyldioxirane (DMDO) and this reaction in combination with a zinc chloride–catalyzed glycosylation step was elaborated by S. Danishefsky's group for solid phase oligosaccharide synthesis.[20]

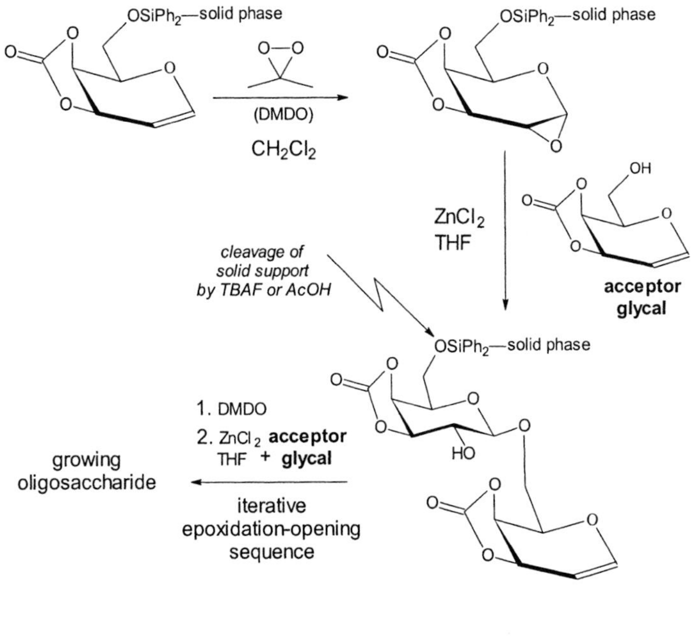

Carbohydrates with terminal exoxyclic double bonds can be used to synthesize carbocyclic compounds. Normal hydroxymercuration of the enol ether group in 6–deoxy–hex–5–enopyranosyl derivatives gives a hemiacetal which equilibrates with the acyclic hex–5–ulose aldehydo–hemicacetals. These are extremely unstable and give the corresponding aldehydo–dicarbonyl organomercury compounds by dehydration. The double activated C–6 nucleophile of the latter compounds, under the conditions of their formation, attack C–1 in aldol fashion to give the carbocyclic hydroxycycloalkanones. This reaction was identified by R. Ferrier and is called the Ferrier (sometimes Ferrier II) reaction.[21]

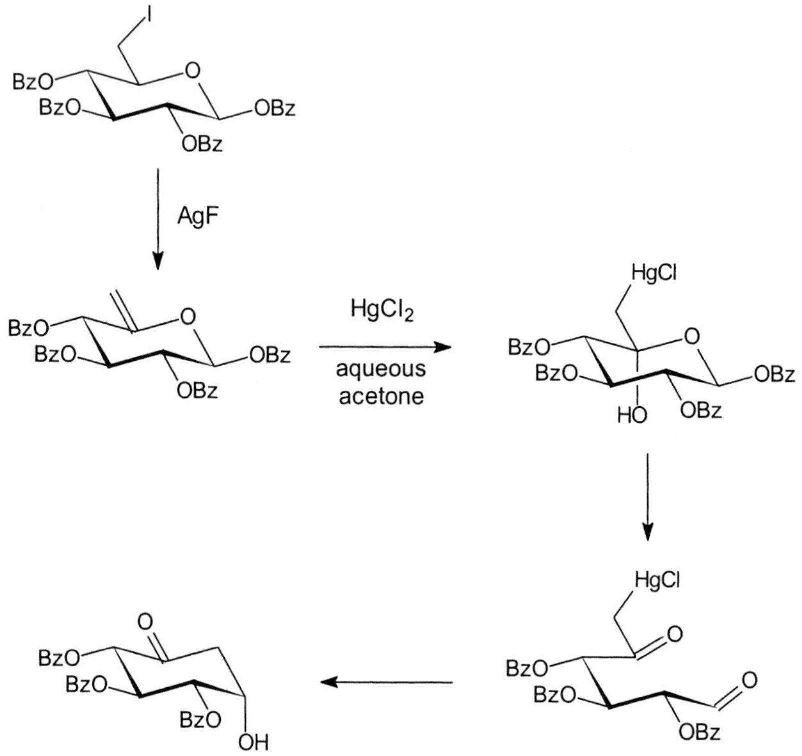

5.5   Epimerizations

Epimerization reactions are used to convert monosaccharides of a given configuration into the desired diastereomer, which would otherwise be difficult to obtain. Different pathways utilizing a variety of reagents are available. For example, an unprotected hydroxyl group in an otherwise blocked monosaccharide can be activated as a sulfonate ester and in turn be substituted by an acetate or benzoate anion to form the acyl esters with inverted configuration at the respective ring carbon atom.

A related method is called the Mitsunobu reaction[22] which employs triphenylphosphine and an azodicarboxylate, mainly diethyl azodicarboxylate (DEAD). In this case, displacement of a particular hydroxyl group occurs *in situ* by activation of the OH–group as alkoxyphosphonium ion intermediate, followed by S_N2 displacement by addition of a nucleophile, an acid in most cases, giving triphenylphosphine oxide and the reduced azodicarboxylate (a hydrazine) as side products.

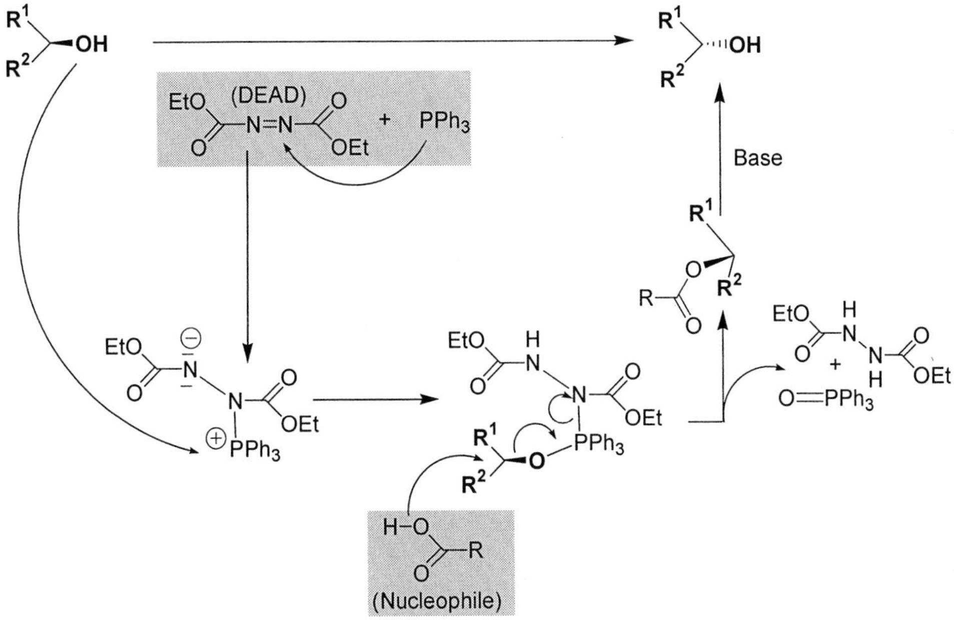

Using benzoic acid in this reaction yields the sugar benzoyl ester with inverted configuration at the treated carbon atom. Thus, the 4–hydroxyl group in an *erythro*–hexenoside can be converted into the *threo*–analog in over 90% yield.

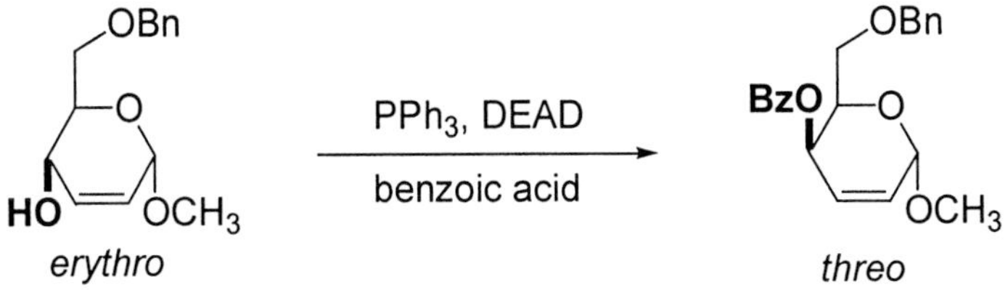

Sometimes the removal of the hydrazine byproduct which is formed by reduction of the azodicarboxylate during the reaction can be problematic. To circumvent this problem, a number of DEAD substitutes have been used in the Mitsunobu reaction. 1,1'–(Azodicarbonyl)dipiperidine seems to be particularly useful as its reduced form is insoluble in organic solvents.

1,1'-(azodicarbonyl)dipiperidine   insoluble in organic solvents

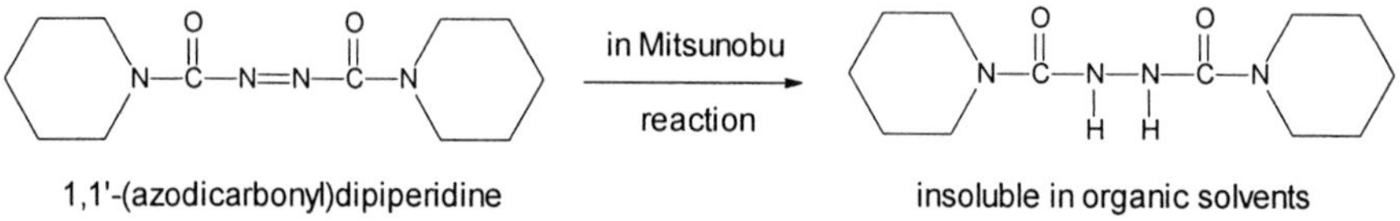

The Mitsunobu reaction can be used for many other purposes, including the formation of deoxyhalo derivatives, thio or anhydro sugars and *O*–glycosides. Modified Mitsunobu procedures use zinc salts instead of acids. The following Mitsunobu reaction with a methyl L–fucoside derivative using ZnI$_2$ led to the 4–deoxy–4–iodo sugar as a result of concomitant acetyl migration.[14]

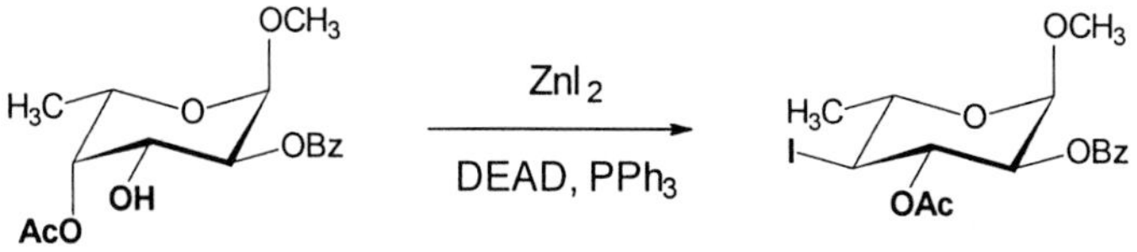

A Mitsunobu reaction has also been employed in the stereoselective synthesis of a branched dideoxy glycoside as shown by P. Rollin and co–workers.[23] In this reaction, the allylic hydroxyl group in the starting hex–2–enopyranoside was reacted with 2–mercaptobenzothiazole with inversion of the configuration. This allows the stereoselective introduction of a *n*–butyl branch at C–2 in a S$_N$2' process, in which the attacking Grignard reagent is anchimerically coordinated by the 2,3–double bond as well as the C=N bond of the axially positioned benzothiazole nucleus. Catalytic reduction of the rearranged product then leads to the C–2–branched dideoxy–ethyl glycoside.

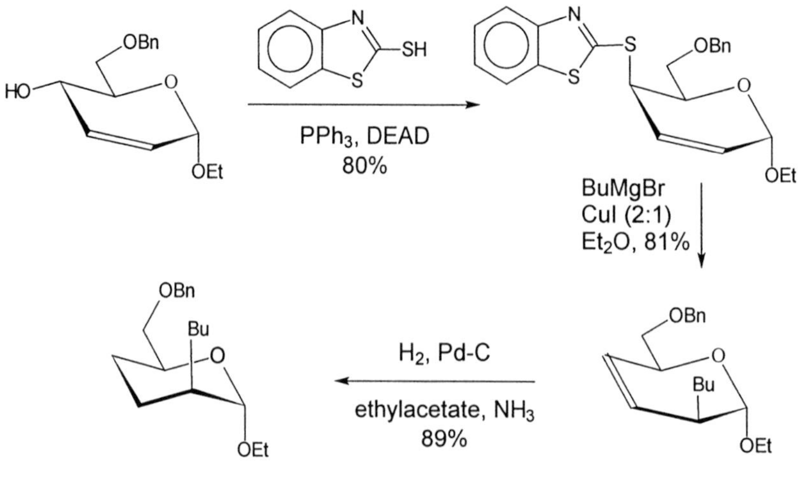

Epimerizations of equatorial hydroxyl groups can often be readily achieved by an oxidation–reduction sequence in which the saccharide is first converted to the respective ulose by oxidation with pyridinium dichromate, for example, and then reduced back to the alcohol. The reduction results in the inverted configuration, since treatment of the ulose with LiAlH$_4$ or L–selectride, respectively, proceeds with high stereoselectivity to produce almost exclusively products with axially disposed hydroxyl groups.[24]

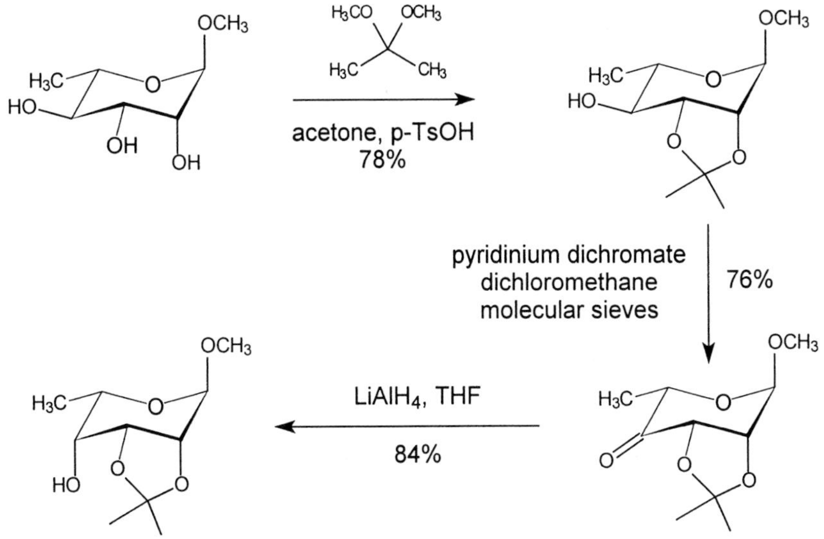

Recently an interesting epimerization reaction has been found by R. Miethchen and co–workers, which employs trichloroacetaldehyde (chloral) and DCC.[25] For this reaction a *cis–trans* sequence of three contiguous hydroxyl groups in the molecule is required. The OH–group in the centre of the three is epimerized in this procedure. The reaction starts with the formation of an equilibrium in which chloral forms a hemiacetal with one of the sugar ring hydroxyl groups and, as a consequence, the acidity of the adjacent OH–groups is enhanced. This allows attack at the carbodiimide carbon, followed by the intramolecular formation of a cyclic imidocarbonate which is in equilibrium with the zwitterion. The latter can rearrange to the epimerized product which carries a carbamoyl function and a trichloroethylidene acetal.

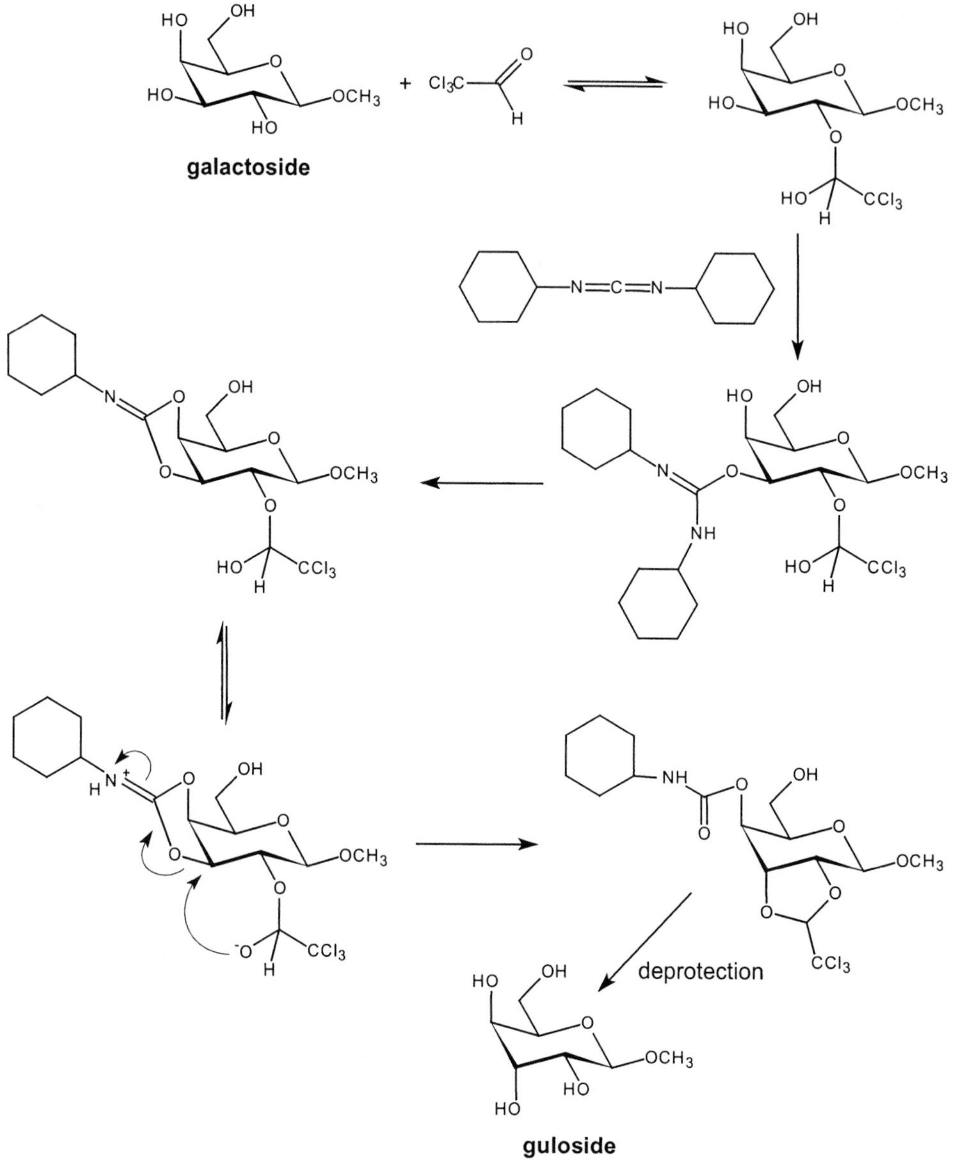

galactoside

guloside

This product can also be regarded as an orthogonally protected building block. The carbamoyl function is easily deprotected in basic medium, however the trichloroethylidene acetal is quite resistant to base and also acid treatment due to its electrondeficient character. It can be cleaved by a two–step procedure, consisting of radical hydrodehalogenation with Bu$_3$SnH, followed by the acidic cleavage of the resulting ethylidene acetal. In this way gulosides for instance, can be obtained from galactosides, and other compounds including inositols have also been synthesized using this procedure.

Configurational inversions at C–5 are attractive because they lead to the production of the generally expensive L–sugars using the D–stereoisomers as starting material. To achieve this, the 6–bromo or 6–iodo D–sugar with the desired stereochemistry is converted by dehalogenation using silver fluoride in pyridine for example, to the corresponding deoxy–hex–5–enopyranoside which contains an exocyclic double bond. Subsequent hydrogenation of the double bond (the C–5 methylene group) leads to the L–configuration of the product.

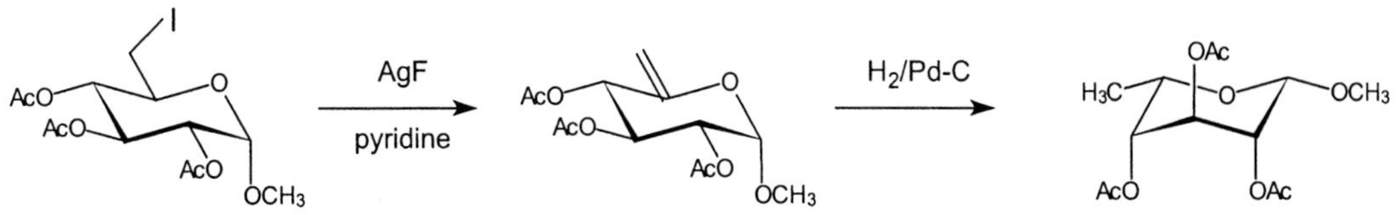

5.6   Carbohydrates as chiral pool materials

Many compounds containing targets carrying several stereocenters can be prepared from carbohydrates in enantiomerically pure form. This is a useful approach to their preparation as pure carbohydrate stereoisomers are obtained from the natural chiral pool in great variety and are often cheap starting materials. However, it requires some experience to recognize which particular arrangement of stereocenters may be deduced from which sugar. The family of aldohexoses provides the following stereochemical sequences of four adjacent stereogenic centers:

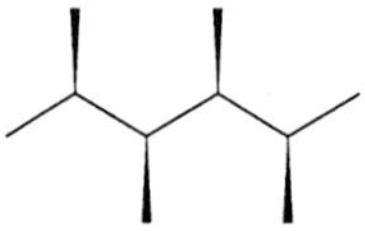

D-ido-

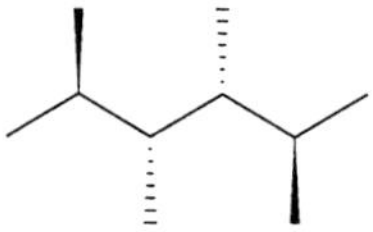

D-manno-

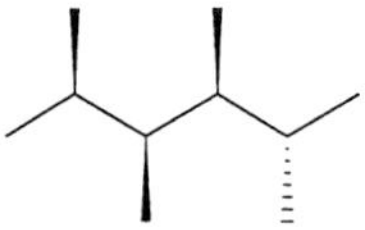

D-gulo-
L-gluco-

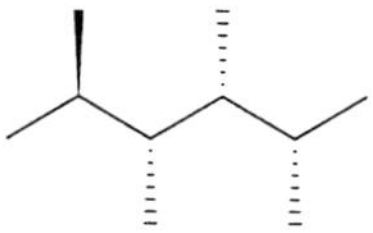

D-gluco-
L-gulo-

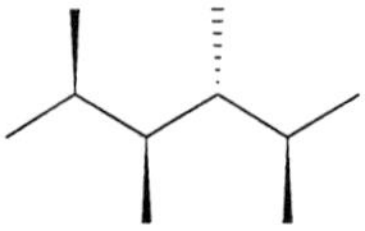

D-altro-
D-talo-

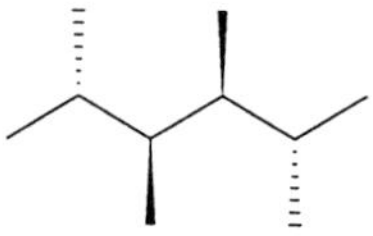

L-manno-

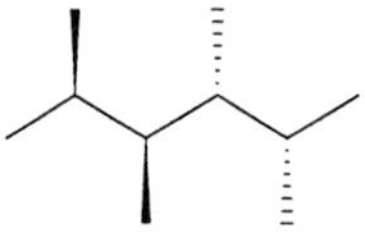

D-galacto-
L-galacto-

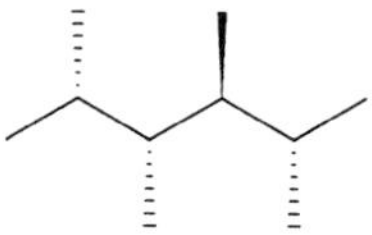

L-altro-
L-talo-

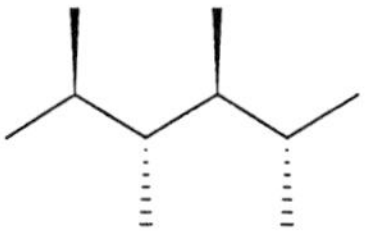

D-allo-
L-allo-

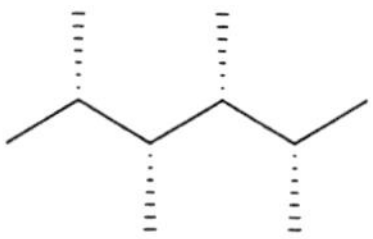

L-ido-

Frequently, carbohydrates contain an excess of stereocenters required for the synthesis of a particular target molecule. In these cases monosaccharides are either split into chiral fragments to be utilized or synthetic pathways are elaborated by which the unnecessary stereogenic centers can be deleted by a short and straightforward procedure. A surprising array of structures can be obtained in this way from saccharide starting materials. Enelactones, for example, can be prepared in a one–pot procedure from (hydroxy)glycal esters, as was shown by F. Lichtenthaler and co–workers.[26] The reaction proceeds by an elimination–oxidation sequence which can be applied to a variety of allylic substrates. It is best carried out by mixing a pre–cooled dichloromethane solution of the (hydroxy)glycal ester and an-hydrous 3–chloroperbenzoic acid followed by the addition of $BF_3 \cdot Et_2O$ and quenching after 15 minutes with saturated sodium hydrogen carbonate. Anhydrous conditions and a reaction temperature of –20°C must be strictly maintained. At higher temperatures oxidative cleavage of the sugar ring occurs.

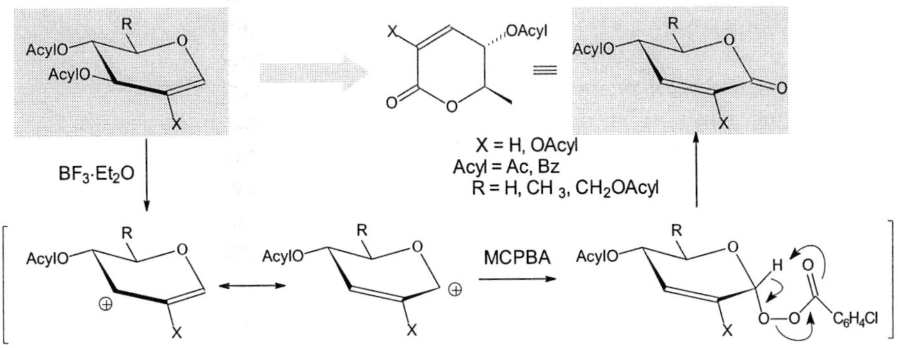

Another interesting example of the use of carbohydrates as chiral building blocks, which also emanated from Lichtenthaler's laboratory, is the stereospecific synthesis of (S,S)–paly-thazine, a natural product synthesized by the salt water invertebrate *Palythoa tuberculosa*, from glucose.[27] In this reaction the glucose–derived 2–O–benzoyl–hydroxyglucal is converted to its oxime, making an enone available after two further steps involving elimination of benzoic acid. This can be converted into the α–ketooxime which, on selective reduction, gives mixed epimeric ketoamines, the dimerization of which results in (S,S)–palythazine.

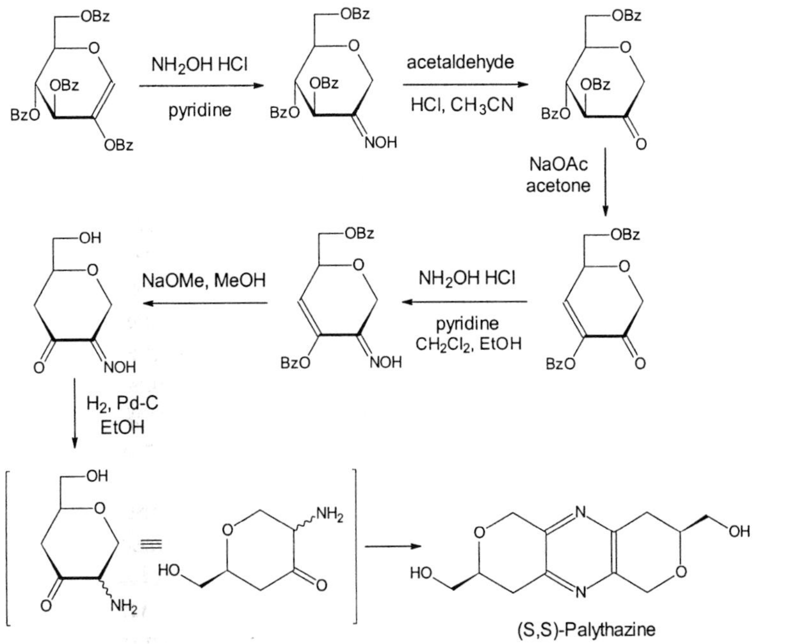

(S,S)-Palythazine

A famous chiral pool synthesis is the preparation of L–ascorbic acid ((R)–5–[(S)–1,2–dihydroxyethyl]–3,4–dihydroxy–5*H*–furan–2–on, vitamin C) from D–glucose. The procedure is based on a microbial oxidation which was first introduced by Reichstein and Grüssner in 1934 and remains the current industrial method of synthesis of L–ascorbic acid with a technical overall yield of 66%. In this reaction the entire carbon chain of D–glucose is inverted in order to produce L–ascorbic acid. First D–glucose is hydrogynated to D–glucitol ('sorbitol'), which is microbially oxidized to L–sorbose by *Acetobacter suboxidans*. After treatment of L–sorbose with acetone in the presence of sulfuric acid, 2,3:4,6–di–*O*–isopropylidene–L–*xylo*–2–hexulofuranose can be isolated. This is oxidized to the corresponding acid, which is heated in water to yield L–*xylo*–2–hexulosonic acid ('2–keto–L–gulonic acid'), from which L–ascorbic acid is obtained by heating in water.

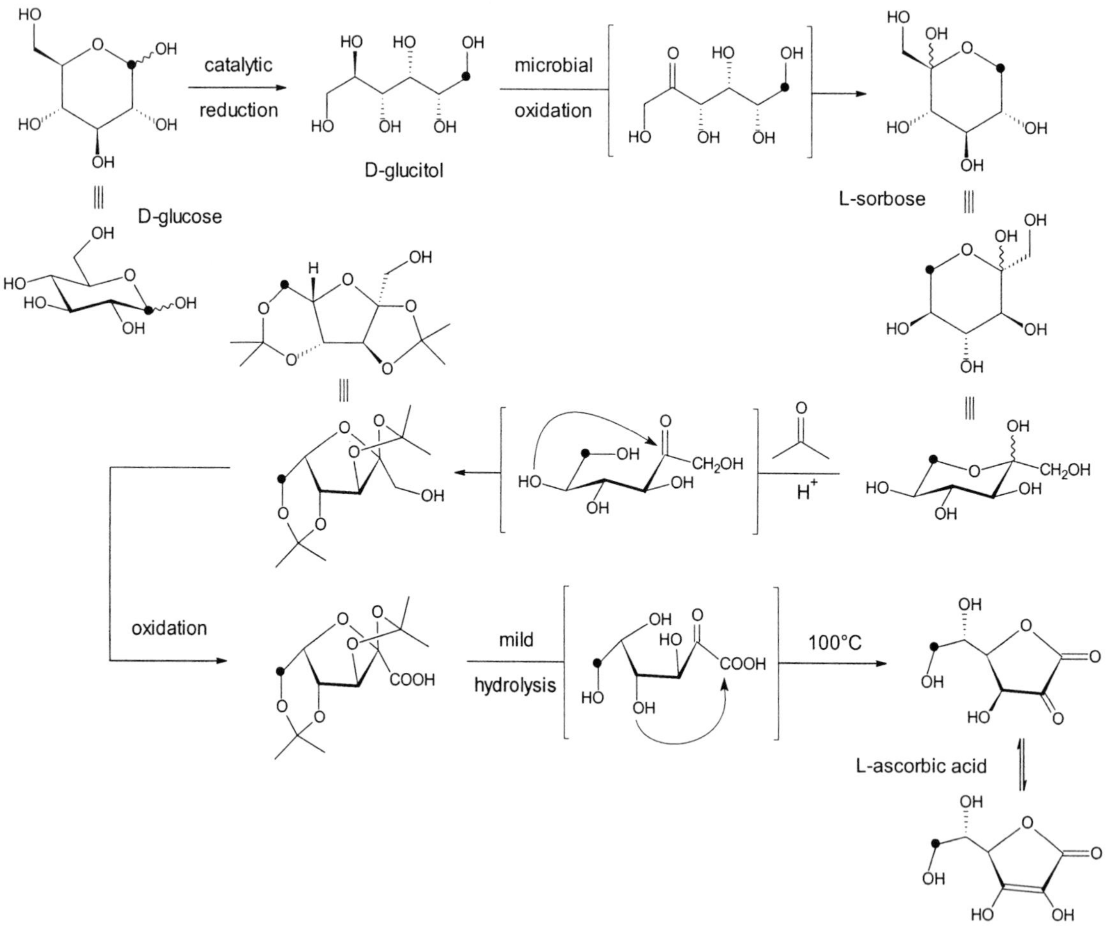

The chemoenzymatic synthesis of L–ascorbic acid from D–glucose. To describe synthetic pathways of this kind, it is feasible to use Mills projections together with the other, more common formulae for carbohydrates. The anomeric carbon atom of D–glucose is tracked (black circles) throughout the synthetic pathway.

The two most attractive features of the Reichstein–Grüssner synthesis are the low cost of D–glucose, and the fact that both the first microbial and the second non–microbiological oxidation steps are carried out on a fully protected intermediate, thus eliminating the possibilities of overoxidation or other side reactions.

Finally an elegant example of the use of sucrose for natural product synthesis is a straightforward chemoenzymatic approach for the preparation of 1–deoxynojirimycin, introduced by A. Stütz and co–workers.[28] This synthesis takes advantage of the poor reactivity to nucleophilic substitution of the 1'–position in sucrose. Therefore, the Appel–type reaction of sucrose with CCl_4 and PPh_3 results in the selective substitution of two of the three primary hydroxyl groups. After substitution of the chlorides by azides, hydrolysis of the glycosidic bond gives 6–deoxy–6–azido–glucose and 6–deoxy–6–azido–fructose. Theses two monosaccharides are separated and the enzyme glucose isomerase is employed to convert the glucose into the respective fructose derivative. The latter is then subjected to hydrogenation resulting in stereoselective intramolecular reductive amination to give 1–deoxynojirimycin.

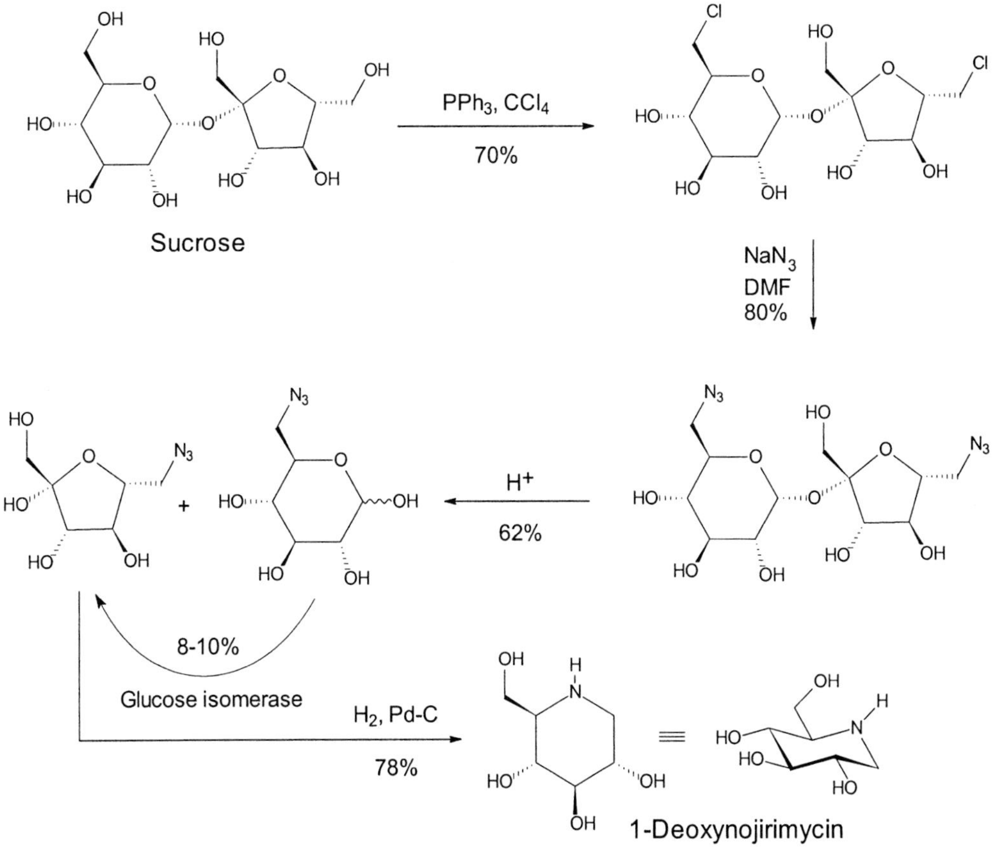

1-Deoxynojirimycin

The product is a 1–deoxy derivative of an azasugar belonging to an important class of glycosidase inhibitors.[29] Azasugars are glycomimetics in which the ring oxygen is substituted with a nitrogen atom. The 1–deoxy derivatives can also be considered as polyhydroxylated piperidines. In their protonated forms, azasugars resemble transition state analogs of the hydrolyzing reaction of glycosidases and are therefore of relevance for mechanistic studies as well as for the treatment of diabetes, for example. Other important members of this class of naturally occurring glycosidase inhibitors are nojirimycin, 1–deoxymanno-nojirimycin, castanosperime and swainsonine.

Nojirimycin   1-Deoxymannonojirimyci

Castanospermine   Swainsonine

References

1.   K. C. Nicolaou, W.-M. Dai (1991) *Angew. Chem. Int. Ed. Engl.* 30, 1387.
2.   J. W. Lown (1993) *Chem. Soc. Rev.* 165.
3.   P. Wang, G.-J. Shen, Y.-F. Wang, Y. Ichikawa, C.-H. Wong (1993) *J. Org. Chem.* 58, 5192.
4.   S. Kötter, U. Krallmann-Wenzel, S. Ehlers, Th. K. Lindhorst (1998) *J. Chem. Soc., Perkin Trans. 1*, 2193.
5.   Th. K. Lindhorst, J. Thiem (1991). *Carbohydr. Res.* **209**, 119.
6.   R. F. Butterworth, S. Hanessian (1971) *Adv. Carbohydr. Chem. Biochem.* 26, 279.
7.   R. U. Lemieux (1963) *Methods Carbohydr. Chem.* 2, 223.
8.   T. Tsuchiya (1990) *Adv. Carbohydr. Chem. Biochem.* 48, 91.
9.   K. Bock, I. Lundt, C. Pedersen (1981) *Carbohydr. Res.* 90, 7.
10.   R. Appel (1975) *Angew. Chem. Int. Ed. Engl.* 14, 801.
11.   P. J. Garegg (1984) *Pure Appl. Chem.* 56, 845.
12.   N. R. Williams (1970) *Adv. Carbohydr. Chem. Biochem.* 25, 109.
13.   D. H. R. Barton, S. W. McCombie (1975) *J. Chem. Soc., Perkin Trans 1*, 1574.
14.   Th. K. Lindhorst, J. Thiem (1990) *Liebigs Ann. Chem.* 1237.
15.   L. Somsak, I. Nemeth (1993) *J. Carbohydr. Chem.* 12, 679.

16. R. U. Lemieux, D. R. Lineback (1965) *Can. J. Chem.* 43, 94.

17. R. U. Lemieux, R. M. Ratcliffe (1979) *Can. J. Chem.* 57, 1244.

18. R. J. Ferrier (1969) *Adv. Carbohydr. Chem. Biochem.* 24, 199.

19. J. S. Brimacombe, L. W. Doner, A. J. Rollins (1972) *J. Chem. Soc., Perkin Trans I*, 1059.

20. P. H. Seeberger, S. J. Danishefsky (1998) *Acc. Chem. Res.* 31, 685.

21. R. J. Ferrier (1979) *J. Chem. Soc., Perkin Trans I*, 1455.

22. O. Mitsunobu (1981) *Synthesis* 1.

23. M. Al Neirabeyeh, P. Rollin (1990) *J. Carbohydr. Chem.* 9, 471.

24. J. Defaye A. Gadelle (1984) *Carbohydr. Res.* 126, 165.

25. R. Miethchen, D. Rentsch, M. Frank (1996) *J. Carbohydr. Chem.* 15, 15.

26. F.W. Lichtenthaler, S. Rönninger, P. Jarglis (1989) *Liebigs Ann. Chem.* 1153.

27. P. Jarglis, F.W. Lichtenthaler (1982) *Angew. Chem. Int. Ed. Engl.* 21, 141.

28. A. de Raadt, A. E. Stütz (1992) *Tetrahedron Lett.* 33, 189.

29. G. Legler (1990) *Adv. Carbohydr. Chem. Biochem.* 48, 319.

6 Structure and biosynthesis of glycoconjugates

Sugars are frequently covalently bound to non–carbohydrate natural products of different kinds. These molecules, which arise from the combination of sugars and other biomolecules, are called glycoconjugates.

The size of glycoconjugates varies from relatively small molecules to large biopolymers. Many of the smaller glycoconjugates possess antibiotic activity. Many pharmaceuticals belong to this category, such as cardiac glycosides, anthracyclines, macrolides, ergot alkaloids, and calicheamycins, to name but a few. These molecules are glycosylated with oligosaccharides of varying complexity, which are important for biological storage and transport and influence the properties of the molecule such as solubility, efficacy, and selectivity among others.

On the other hand, even more complex carbohydrates are linked to proteins and lipids producing a large number of different glycoconjugates called glycoproteins, proteoglycans, glycolipids and GPI-anchors, respectively. These glycoconjugates belong to the most important group of biomolecules in the cell. Their carbohydrate moieties comprise an enormous structural variety and this is used by nature to store biological information which is required in cell adhesion processes. Understanding the biological functions of the complex carbohydrate chains present in glycoconjugates has thus become one of the major topics of a growing field of research, called glycobiology. Before their functions came into the focus of carbohydrate research, the structures and biosynthesis of glycoconjugates were first elucidated.

6.1 Structural diversity of oligosaccharides

Carbohydrates are unique in the complexity of their structures. In contrast to the other two major classes of biologically important biopolymers, proteins and nucleic acids, oligo– and polysaccharides are built up of monomers which have more than two functional groups participating in an oligomerization reaction. In other words, in a sugar residue one or more of several different hydroxyl groups can be glycosylated, thus also allowing the formation of branched structures. Furthermore, the glycosidic linkage can lead to one of two different stereoisomers, the α– or the β–glycoside.

Consequently, many more constitutional stereoisomers can be constructed from monosaccharides than from amino acids or nucleotides from which only linear oligomers can be constructed (Table 6–1).

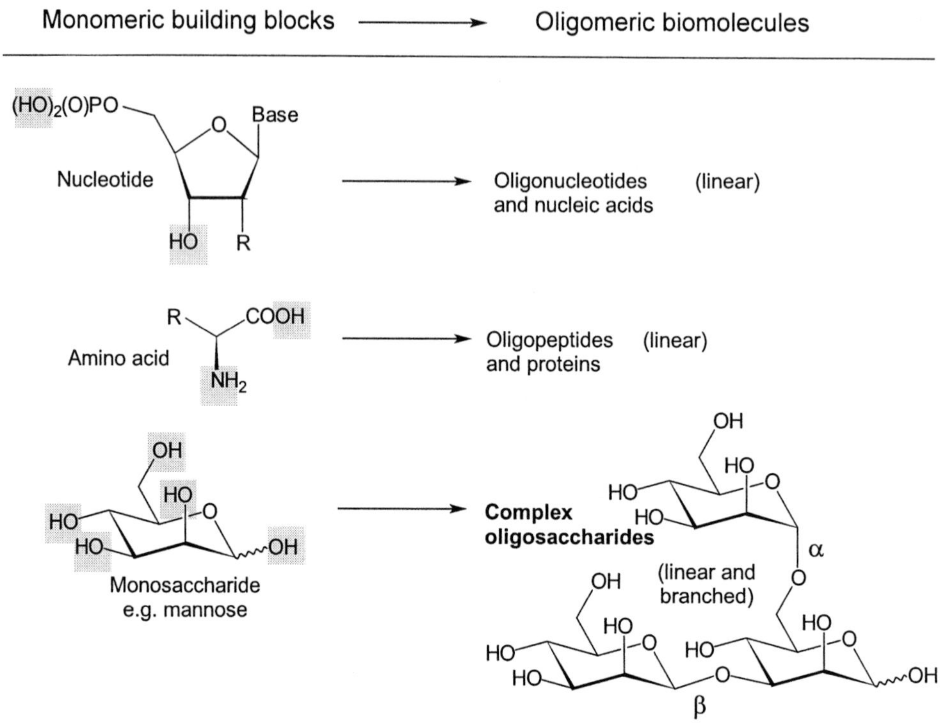

Monomeric building blocks ⟶ Oligomeric biomolecules

Oligomerization of nucleotides or amino acids, respectively, leads to linear oligomers and polymers. In contrast, monosaccharides can be linked together to form a variety of linear as well as branched oligosaccharides (cf. Table 6–1). Functional groups which react in the oligomerization reactions are highlighted in gray.

Table 6–1. Numbers of theoretically possible homo– and heterooligomeric stereoisomers derived from oligonucleotides and peptides on the one hand and oligosaccharides on the other.

Oligomer	Composition	Possible oligopeptides and oligonucleotides	Possible oligosaccharides
Dimer	AA / AB	1 / 2	11 / 20
Trimer	AAA / ABC	1 / 6	120 / 720
Tetramer	AAAA / ABCD	1 / 24	1424 / 34560
Pentamer	AAAAA / ABCDE	1 / 120	17872 / 2144640

The extremely large structural diversity of oligosaccharides is indeed utilized in nature in molecular recognition processes. Although oligosaccharides do not encode biological information in form of their monomer sequences in the sense of a biological code as in oligonucleotide sequences, which carry the genetic code, they do contain biological information in form of their three–dimensional structures and this is of importance for cell–cell adhesion and cell communication. In these processes the carbohydrate structures of glycoconjugates function as ligands for specialized proteins, called lectins and selectins, which recognize specific structural arrays of the saccharide moieties. In these ligand–receptor interactions non–covalent carbohydrate–protein complexes are formed and this may eventually trigger a cascade of further biological events. The elucidation of the molecular details of recognition processes, in which glycoconjugate saccharides are involved, has as yet only just begun. One of the early questions to be asked is, what do biologically important oligosaccharides look like and how are they presented to their receptors in order to fulfill their function as biological markers and signaling molecules?

6.2   Structures of glycoconjugates

Glycoconjugates are ubiquitously found in nature. Proteins, such as enzymes, antibodies, hormones, cytokines and receptor proteins, are glycosylated with their carbohydrate content varying from 1% (in collagen) to 99% (in glycogen). The carbohydrate portions of proteins can alter the biological and physicochemical properties of the conjugate such as their stability against proteases or the activity of enzymes and they can direct the folding of proteins toward certain three–dimensional structures.

The oligosaccharide moieties in glycoconjugates typically consist of up to 20 monomers. They may also be much larger or smaller, as in the case of proteins which are only glycosylated with mono– or disaccharides. Although there is a vast array of diversity in carbohydrate structures, only a relatively small set of monosaccharide building blocks are used in the biosynthesis of glycoconjugate oligosaccharides out of the large variety of available monosaccharide stereoisomers. Therefore, some general oligosaccharide patterns and certain linking stereochemistries are typically found in glycoconjugates. In spite of these limitations, the stereochemical variety in which oligosaccharides can occur in glycoconjugates is almost infinite. The variety of the structures that can occur can even be increased by derivatizations of the sugar rings such as *O*–methylation, *O*–acetylation, *O*–sulphatation, *O*–phosphorylation or oxidation.

Glycoproteins are found in soluble form, for example in the blood, in the cytosol or in subcellular organelles. Furthermore, they are basic constituents of all cell membranes. In eukaryotic cells they are integrated into the lipid bilayer, so that the oligosaccharide moieties are exposed to the extracellular side of the membrane. There they form a carbohydrate coat which is called a 'glycocalyx'. The glycocalyx oligosaccharides are major components of the eukaryotic cell surface and can expand to form a layer of up to 140 nm in depth.

D-Glucose (Glc)   D-Mannose (Man)   D-Galactose (Gal)
(enantiomer also found)   D-Fructose (Fru)

D-Glucuronic acid   D-Mannuronic acid   D-Galacturonic acid   L-Iduronic acid
(GlcA)   (ManA)   (GalA)   (IduA)

D-Xylose (Xyl)   D-Ribose (Rib)   L-Arabinose (Ara)
(enantiomer also found)   2-Deoxy-D-ribose
(dRib)

N-Acetyl-D-glucosamine   N-Acetyl-D-galactosamine   L-Fucose (Fuc)   L-Rhamnose (Rha)
(GlcNAc)   (GalNAc)

3-Deoxy-α-D-*manno*-
oct-2-ulopyranosonic acid
(Kdo)   5-Acetamido-3,5-dideoxy-D - *glycero*-
α-D-*galacto*-non-2-ulopyranosonic acid
N-Acetyl-neuraminic acid (Neu5Ac)   N-Acetyl-muraminic acid
(MurAc)

Structures and names of the most important monosaccharide constituents in glycoconjugates with their generally used abbreviations.

Table 6–2. The main monosaccharide elements used in the biosynthesis of glycoconjugate oligosaccharides and their stereochemistries.

Monosaccharide	Glycosidic linkage	Occurence in glycoconjugates
D–Glucose	α or β	Mainly in collagen
D–Galactose	α or β	Ubiquitous
D–Mannose	α or β	Ubiquitous
N–Acetyl–D–glucosamine	α or β	Ubiquitous
N–Acetyl–D–galactosamine	α or β	Hardly in plants
N–Acetyl–neuraminic acid	α	Only in higher invertebrates and in vertebrates
L–Fucose	α	Ubiquitous
D–Xylose	β	In plants and proteoglycans
3–Deoxy–D–*manno*–octulosonic acid	α	In lipopolysaccharides

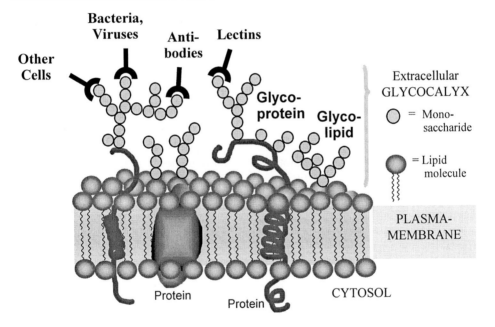

Carbohydrates in glycolipids and glycoproteins are exposed on cell surfaces like molecular antennae. There they form ligands, which can be recognized by a number of different molecules such as lectins and selectins, enzymes, hormones, toxins or antibodies which use oligosaccharides as antigenic markers to distinguish 'own' from 'other'. Interestingly, microbes also use the oligosaccharide surface of potential host cells as attachment points with the help of own lectins. These often undesired carbohydrate–protein interactions can lead to the firm adhesion of microbes to the host cell surface, and this event can be followed by invasion of the cell.

The carbohydrate coat of a particular cell is characteristic for its type and for its developmental, physiological and even pathological status. In embryonal cells or in cancer cells, for example, the glyco–coat varies significantly, compared with adult or healthy cells respectively. Thus certain oligosaccharides can be causally associated with degenerative cell growth and can be used in cancer diagnostics as so–called tumor–associated antigens.

6.2.1   Structures of glycoproteins

Proteins may contain one or several oligosaccharide side chains. The three most important chemical linkage types, in which carbohydrates are covalently bound to proteins are the following:

(i)   *N–glycosidic*: The resulting conjugates are called *N–*glycoproteins or *N–*glycans. In *N–*glycoproteins the oligosaccharide moiety is *always* bound to the side chain of an asparagine amino acid of the protein moiety via an *N–*glycosidic linkage to *N–*acetylglucosamine, which forms the non–reducing end in *all N–*glycans. All *N–*glycoproteins have a particular internal pentasaccharidic region in common.

(ii)   *O–glycosidic*: The resulting conjugates are called *O–*glycoproteins or *O–*glycans. In *O–*glycoproteins the reducing end of the oligosaccharide chain is *O–*glycosidically linked to an OH–group in the side chain of an amino acid of the protein moiety, which is mostly serine or threonine. The nature of the monosaccharide moiety which is attached to the peptide varies between *O–*glycoproteins. A uniform internal core region is not observed.

(iii) via **ethanolamine phosphate**: This linkage type between the protein and carbohydrate moieties in glycoconjugates occurs in glycosylphosphatidylinositols, called GPI–anchors, which anchor proteins in cell membranes.

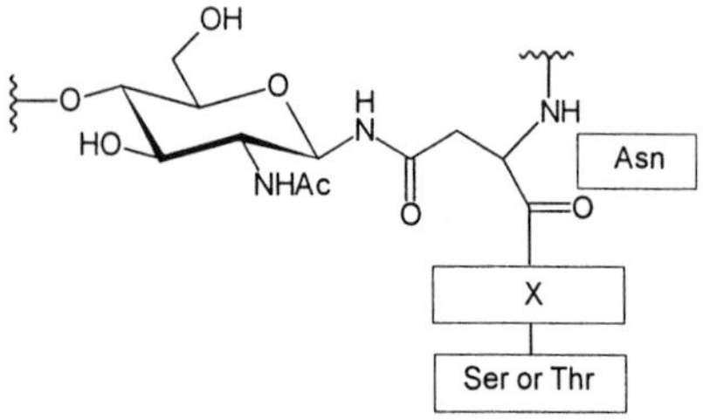

N-glycosidic 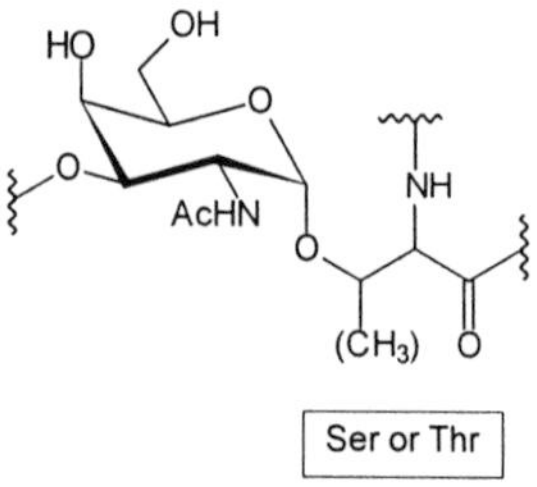   O-glycosidic 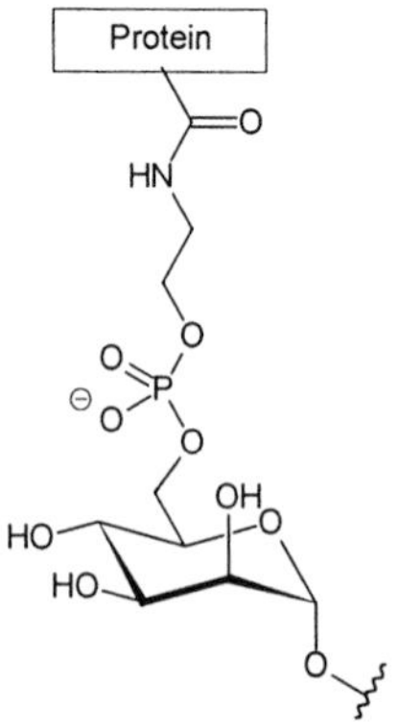   *via* ethanolamine phosphate

N–Glycans

All *N*–glycoproteins share a peptide–linked pentasaccharide fragment, which is called the core region. It consists of a branched structure Man–(α1,6)[Man–(α1,3)]Man–(β1,4)–GlcNAc–(β1,4)GlcNAc, with the terminal GlcNAc *N*–glycosidically linked to an aspar-agine residue of the peptide chain in β–mode.

Branching out from this uniform core are oligosaccharide chains of diverse structural va-riety achieved by the attachment of different saccharides, leading to multiple branched or unbranched structures. Certain commonalities can be identified in *N*–glycans and based on this finding, they have been classified into three types,

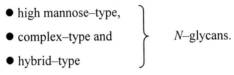

- high mannose–type,
- complex–type and   *N*–glycans.
- hybrid–type

These common characteristics can be explained on the basis of the biosynthesis of *N*–glycoproteins, in which all oligosaccharide chains are derived from one and the same lipid–bound precursor saccharide, which contains the core region common to all *N*–glycans. It remains intact in all the steps of the biosynthetic pathway in which the glycoprotein oligo-saccharides are processed to the three different *N*–glycan types each in a large number of variations. Thus, *N*–glycan chains differ in number and type of monosaccharides linked to-gether, as well as in the number of side chains which are formed by branching and which are called antennae. Moreover, sugar residues in glycoconjugates can be modified by various functionalities such as sulfate, phosphate or carboxy groups and others.

The simplest *N*–glycans are high mannose–type oligosaccharides which contain only α–mannosyl residues bound to the branching *N*–glycoprotein core. In complex type *N*–glycans the α–mannose residues forming the bisecting core are elongated with *N*–acetylglucosamine residues and *N*–acetyllactosamine (Galβ1→4GlcNAc) disaccharide moieties. Numerous variations arise from differences in the number of *N*–acetylglucosamine residues and their attachment positions, resulting in multiple–branched structures, known as bi–, tri–, tetra, and pentaantennary. Further diversity results from variations in the side chains which can terminate in galactose, fucose or neuraminic acid residues. Moreover, the bisecting core region can be altered by glycosylation with L–fucose and/or *N*–acetylglucosamine. In com-plex type *N*–glycoproteins an *N*–acetyllactosamine disaccharide moiety of the structure [3–Gal–(β–1,4)–GlcNAc–β] is often found as a repeating unit, which can be attached in groups of two to up to approximately 50. These *N*–LacNAc–oligomers are referred to as poly–*N*–acetyllactosamino glycans. They are typically represented in tetraantennary oligosaccharide structures. Hybrid–type oligosaccharides possess structural elements of both of the other two classes. They contain more than three mannose moieties but also *N*–acetyllactosamine side chains. Their branches often terminate with *N*–acetylneuraminic acid.

In spite of the observed commonalities of *N*–glycans, their structural diversity is almost unlimited, in particular that of complex type–oligosaccharides.

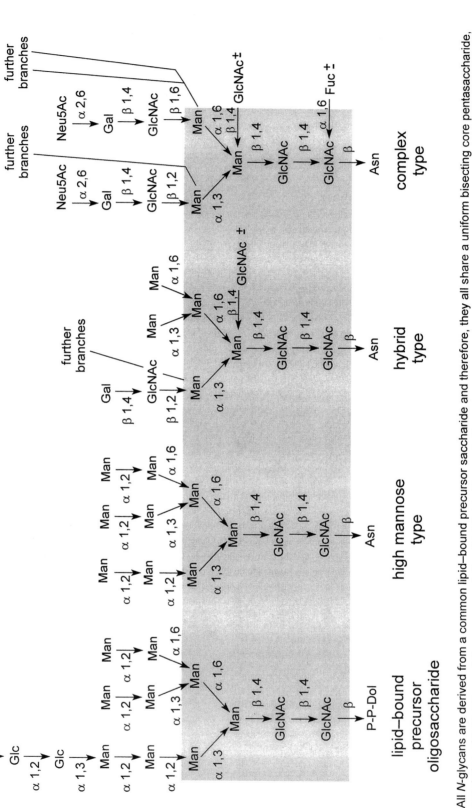

All *N*-glycans are derived from a common lipid–bound precursor saccharide and therefore, they all share a uniform bisecting core pentasaccharide, consisting of 2 GlcNAc and 3 mannose moieties (highlighted in gray). Complex type–glycans are the most highly branched oligosaccharides, which represent up to pentaantennary structures. The core region of complex–type oligosaccharides can be further modified by the attachment of an α–fucosyl or/and a β-*N*-acetylglucosaminyl residue.

O–Glycans

This group of glycoproteins was initially found in a mucus substance called mucin and the component oligosaccharides are therefore known as mucin–type glycans. So far they have been less intensively investigated than *N*–glycan structures, while their structural heterogeneity seems to be even greater. Unlike *N*–glycans, there is no common core region in *O*–glycoproteins. The monosaccharide at the reducing end of the glycan moiety is *O*–glycosidically linked to the hydroxyl group of a serine or threonine amino acid side chain in most cases. In mammalian tissue this terminal sugar is often found to be α–*N*–acetylgalactosamine. The attempt to classify mucin–type oligosaccharide structures has led to the identification of six major groups of *O*–glycans with a peptide–bound GalNAc residue, called core class I to VI.

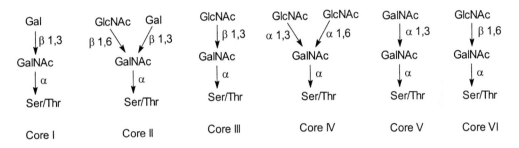

In addition to the above–mentioned core classes, other structural patterns exist within the *O*–glycans. For example, glycosylation may occur with an *N*–acetylglucosamine or fucose residue rather than GalNAc.

The functions of *O*–glycoproteins are still under investigation. They seem to be more involved in the stabilization and protection of protein structures rather than functioning as signaling molecules in cell communication processes.

GPI–Anchors

Glycosylphosphatidylinositol molecules, called GPI–anchors, were originally found in parasites but recent research has increasingly shown their presence in various mammalian glycoproteins. They have a structural element in common which is peptide–linked to the *C*–terminus of their protein part. It has the structure

$$\text{ethanolamine–PO}_4\text{–6Man}\alpha1\rightarrow2\text{Man}\alpha1\rightarrow6\text{Man}\alpha1\rightarrow4\text{GlcNH}_2\alpha1$$
$$\downarrow$$
$$\text{Lipid–PO}_4\text{–1–inositol–}myo\text{–6}$$

Depending on the species and the type of tissues in which they occur, the basic structure of GPI–anchors is modified by many different oligosaccharide and lipid chains, as well as by other modifications. The lipid moiety, which usually consists of a diacyl glycerol or a ceramide moiety, anchors the molecule in the cell membrane such that the GPI protein part is exposed to the extracellular side of the plasma membrane.

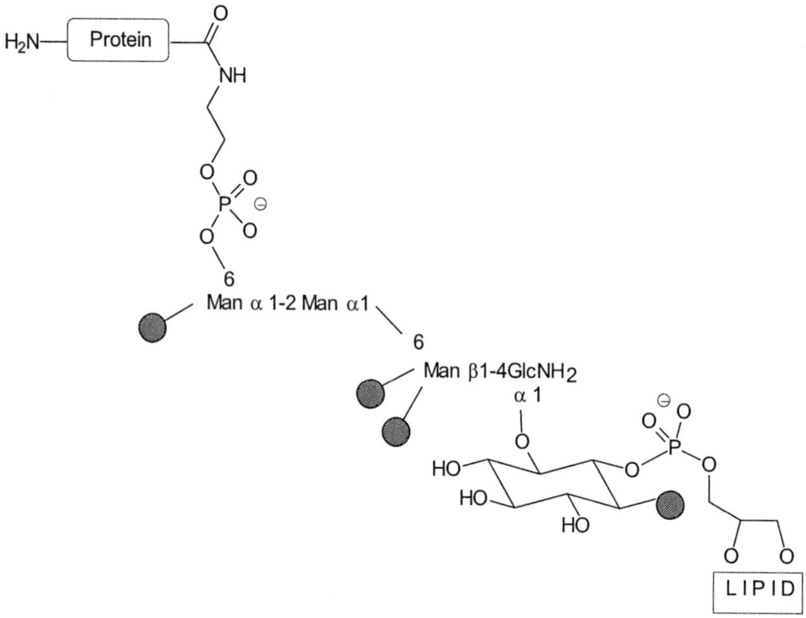

The basic structure of glycosylphosphatidylinositols (GPI–anchors) consists of protein, oligosaccharide, and lipid portions and an inositol moiety. The filled circles indicate variable sites of the molecule, where further oligosaccharide, lipid, phosphate, or ethanolamine moieties can be attached.

Proteoglycans

Proteoglycans are a special class of glycoproteins. They differ substantially from other glycoproteins in that they contain large protein–bound polysaccharides. These polysaccharides, which are also called glycosaminoglycans, are of polyanionic character and consist mainly of repeating disaccharidic subunits, which are often *O*–sulfated to varying extents. Depending on their composition, glycosaminoglycans are known as hyaluronic acid (hyaluronan), chondroitin sulfate, dermatan sulfate, heparin, heparan sulfate, or keratan sulfate, the latter lacking uronic acids (cf. chapter 2.3).

Table 6–3. Composition of glycosaminoglycans.

Basic disaccharide unit	Common names
GlcNAcβ1,4–GlcAβ1,3	Hyaluronic acid
GalNAc(6 or 4–SO$_3^-$)β1,4–GlcAβ1,3	Chondroitin sulfate
GalNAc(4–SO$_3^-$)β1,4–L–IduAα1,3	Dermatan sulfate
GlcNSO$_3$(6–SO$_3^-$)α1,4–L–IduAα1,4	Heparin
GlcNSO$_3$β1,4–L–IduAα1,4	Heparan sulfate
GlcNAc(6–SO$_3^-$)β1,3Gal(6–SO$_3^-$)β1,4	Keratan sulfate

6.2.2 Structures of glycolipids

Complex lipids are, as all lipids, amphiphilic molecules containing a hydrophilic and a lipophilic part. The lipophilic part consists of either 1,2–di–*O*–diacylglycerol or *N*–acylsphingosin. The hydrophilic element of the lipid is provided either by a phosphate group in the case of glycerophospholipids or sphingophospholipids, or by a carbohydrate moiety as in glycoglycerolipids and in glycosphingolipids, which are found in plants and mammals, respectively. Glycosphingolipids are important constituents of eukaryotic cell membranes.

The hydrophobic entity in glycosphingolipids is *N*–acylsphingosin and is called ceramide. It anchors the molecule in the outer half of the plasma membrane such that its oligosaccharide component is exterior the cell.

Several hundred different glycosphingolipid structures are known. The nature of both the hydrophobic chains in the ceramide moiety and in particular their carbohydrate content vary considerably. The simplest glycosphingolipids are called cerebrosides and contain only one monosaccharide per ceramide, such as glucosyl cerebrosides and galactosyl cerebrosides. The latter are widely distributed in the plasma membranes of neuronal cells. Several galactosyl cerebrosides are sulfated at the galactose moiety and are therefore known as sulfatides.

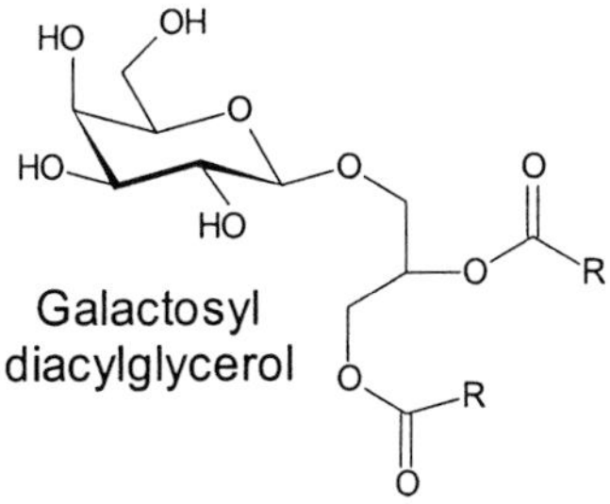

Galactosyl diacylglycerol

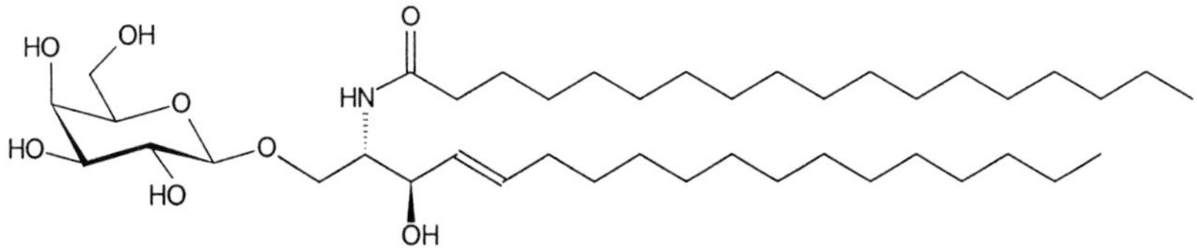

Galactosyl cerebroside

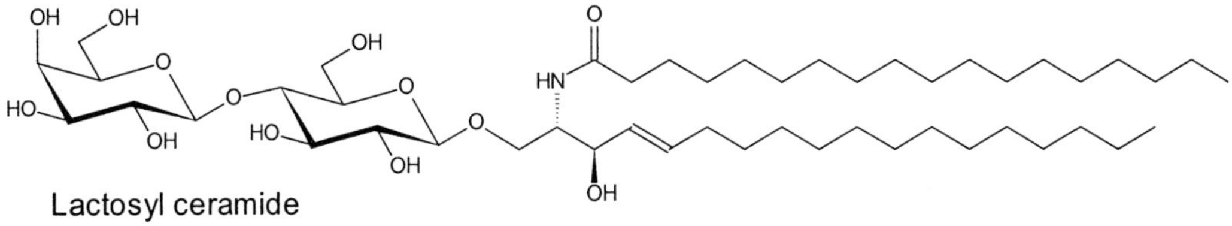

Lactosyl ceramide

Examples of the structures of glycoglycerolipids and glycosphingolipids.

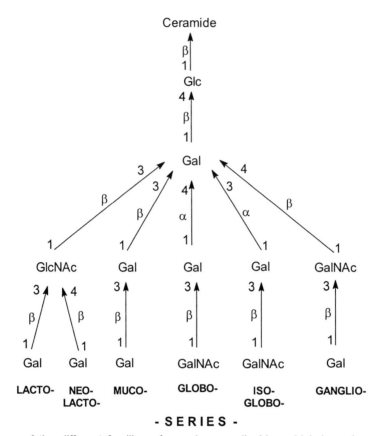

Structures of the different families of complex gangliosides which have in common a β–lactosyl moiety in the ceramide.

Structure of the ganglioside GD1a.

More complex glycosphingolipids are called gangliosides (neutral glycosphingolipids, containing neutral oligosaccharides are sometimes called globosides). All gangliosides have in common a β–lactosyl moiety in the ceramide, as the first two glycosylation steps in the biosynthesis of gangliosides are the same for all members of this group. Lactosylceramides are further modified to produce a wide variety of structures glycosylated in more complex ways. These are classified into different series depending on which features in their carbohydrate patterns they have in common.

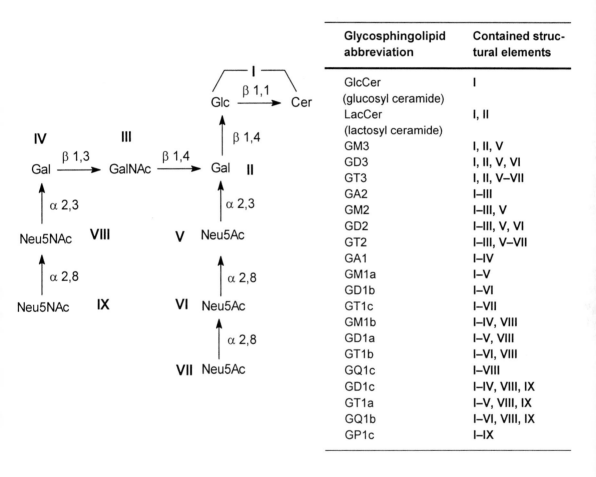

Glycosphingolipid abbreviation	Contained structural elements
GlcCer (glucosyl ceramide)	I
LacCer (lactosyl ceramide)	I, II
GM3	I, II, V
GD3	I, II, V, VI
GT3	I, II, V–VII
GA2	I–III
GM2	I–III, V
GD2	I–III, V, VI
GT2	I–III, V–VII
GA1	I–IV
GM1a	I–V
GD1b	I–VI
GT1c	I–VII
GM1b	I–IV, VIII
GD1a	I–V, VIII
GT1b	I–VI, VIII
GQ1c	I–VIII
GD1c	I–IV, VIII, IX
GT1a	I–V, VIII, IX
GQ1b	I–VI, VIII, IX
GP1c	I–IX

Glycosphingolipids are named using a series of abbreviations. The letter following 'G' indicates the number of neuraminic acid residues (M for one, D for two, T for three, Q for four, and P for five) and the index number represents the result of [5 minus (the number of neutral monosaccharide units)]. Since the nomenclature of glycosphingolipid molecules is not self–evident, a scheme is provided to simplify the identification of these.

Lipopolysaccharides

Lipopolysaccharide (LPS), also called endotoxin, is located in the outer membrane of Gram–negative bacteria. LPS participates in many physiological processes and plays a key role in the pathogenesis and manifestation of Gram–negative infection in general and septic shock in particular.

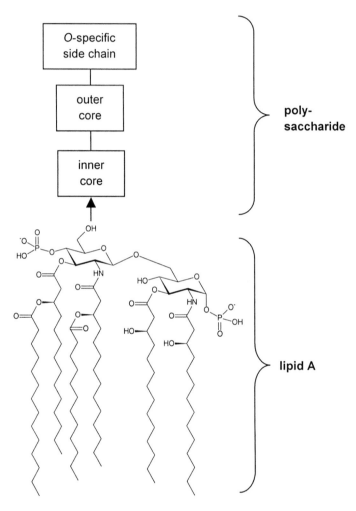

Principal structural regions of LPS molecules. The structure of the lipid A is given for a well–known *Escherichia coli* strain (Re mutant strain F515). The lipid A of this *E. coli* strain is composed of a 2–amino–2–deoxy–6–*O*–(2–amino–2–deoxy–β–D–glucopyranosyl)–α–D–glucopyranose disaccharide [β–D–GlcNAc–(1→6)–α–D–GlcNAc], which carries two phosphate groups. The 6'–position is only unsubstituted in free lipid A, in LPS it is linked to a Kdo portion of the polysaccharide chain.

Lipopolysaccharides are amphiphilic macromolecules, containing three distinct regions which can be distinguished genetically and chemically as well as antigenically. These regions are named the *O*–specific side chain, the core–oligosaccharide, and lipid A. The *O*–specific side chain is a heteropolysaccharide, composed of up to 60 repeating units, which are themselves made up of two to six sugar monomers. It is specific for a bacterial species and has an enormous structural variety. The core oligosaccharide comprises of two regions, the outer and the inner core. These core regions have less structural variability than the *O*–specific chain. The outer core region of enterobacterial LPS contains hexoses such as D–galactose, D–glucose, GlcNAc and GalNAc, while the inner core region is mainly composed of rare long chain carbohydrates such as L–glycero–D–manno–heptose or Kdo, which are often functionalized with phosphate or 2–aminoethyl(pyro)phosphate. The lipid A region consists of the covalently bound lipid component serving to anchor LPS to the bacterial membrane. It is the least variable component of LPS. The primary structure of lipid A in various Gram–negative bacteria has been the subject of intensive research and has been elucidated in detail in many cases.

6.3   Biosynthesis of oligosaccharides

The biosynthesis of glycoconjugate oligosaccharides of mammalian cells is catalyzed by glycosyltransferases, called Leloir enzymes after the scientist who discovered them. They use nucleosidediphosphate–sugars as activated substrates for the stereo– and regiospecific transfer of the respective monosaccharide onto an acceptor saccharide. Glycosyl transfer is facilitated by the superior leaving group properties of nucleotide diphosphates.

The synthesis of the glycosyltransferase substrates starts with a monosaccharide, which is stereospecifically converted by a kinase to the respective glycosylphosphate. The latter is then converted to a nucleosidediphosphate sugar, as a result of pyrophosphorylase catalysis. This reaction requires the respective nucleosidetriphosphate as co–substrate.

The biosynthesis of the activated form of neuraminic acid, CMP–Neu5Ac, differs from that in other cases, as it is directly synthesized from *N*–acetylneuraminic acid and CTP, catalyzed by CMP–Neu5Ac synthase; no kinase–catalyzed intermediate step is required.

Biosynthesis of CMP–Neu5Ac.

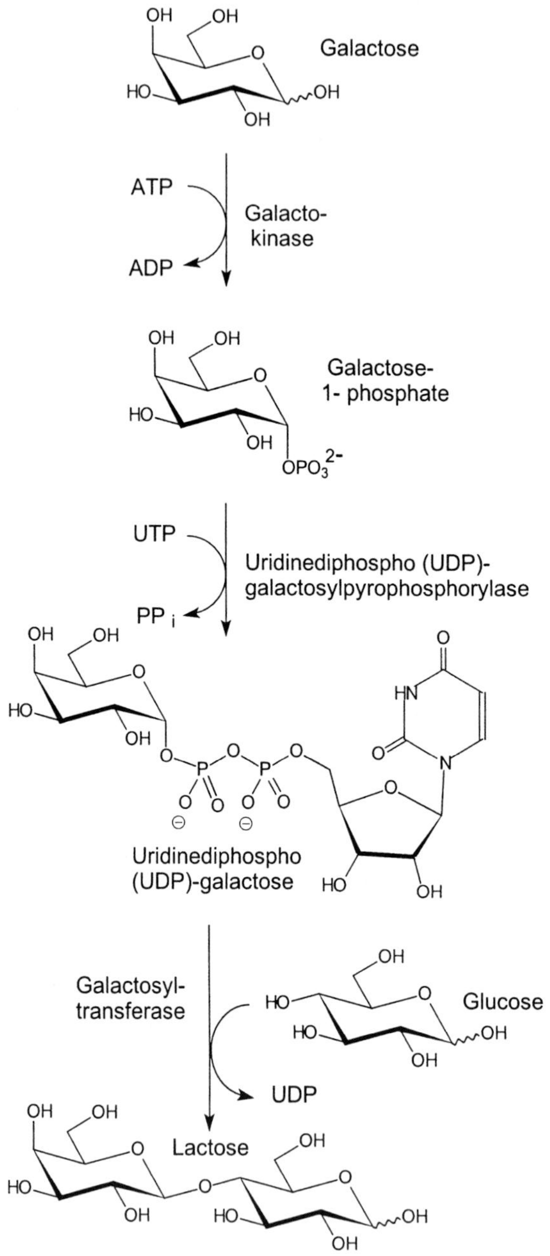

Biosynthesis of UDP–galactose and the enzyme–catalyzed transfer of a galactose moiety to glucose to form the disaccharide lactose.

For the synthesis of most oligosaccharides, Leloir transferases utilize a set of activated donor molecules, listed in Table 6–3.

Table 6–3. Activated saccharides (XDP–sugars) used as substrates for Leloir transferases. *N*–Acetylneuraminic acid is an exceptional because it is the only monosaccharide which is not activated as a nucleosidediphosphate but as the monophosphate (CMP–Neu5Ac).

XDP–sugar	Used for the transfer of
Uridine–5'–diphosphoglucose (UDP–Glc)	Glucose
Uridine–5'–diphosphogalactose (UDP–Gal)	Galactose
Uridine–5'–diphospho–*N*–acetylgalactosamine (UDP–GalNAc)	GalNAc
Uridine–5'–diphospho–N–acetylglucosamine (UDP–GlcNAc)	GlcNAc
Uridine–5'–diphosphoglucuronic acid (UDP–GlcA)	GlcA
Guanosine–5'–diphosphofucose (GDP–Fuc)	L–Fucose
Guanosine–5'–diphosphomannose (GDP–Man)	Mannose
Cytidine–5'–monophospho– *N*–acetylneuraminic acid (CMP–Neu5Ac)	Neu5Ac

Other than the biosynthesis of proteins, which is directly dependent on the genetic code, the structure of oligosaccharides is determined by the action of enzymes, and therefore oligosaccharides can be called secondary gene products. Consequently, glycosyltransferases evolved as very specific enzymes, with only one particular transferase being able to catalyze one particular glycosidic linkage. This 'one enzyme–one linkage' rule was found to be almost a dogma of glycobiology and only very few exceptions have been found. However, as enzymatic glycosylations are also dependent on the actual reaction conditions, this results in small differences in the oligosaccharide structures biosynthesized which leads to heterogeneity among the oligosaccharide patterns of cells, known as microheterogeneity.

6.4   Biosynthesis of *N*–glycoproteins

The biosynthesis of *N*–glycans can be divided into three parts:

(i)   The phosphodolichol cycle occurring in the rough endoplasmic reticulum: A lipid–bound oligosaccharide is synthesized, which serves as the precursor for all the various oligosaccharides expressed in *N*–glycoproteins. It already contains the *N*–glycan core structure.

(ii)   The precursor oligosaccharide is then transferred from the lipid–bound form to an asparagine side chain of a polypeptide chain, growing at the endoplasmic reticulum. This process is catalyzed by an enzyme called oligosaccharyltransferase. The site on the polypeptide to which the saccharide is attached contains a consensus region of the amino acid sequence Asn–X–Ser/Thr, where X can be any amino acid, except proline.

(iii)   The so–called trimming and processing of the oligosaccharide then takes place in the medial and trans–compartments of the Golgi apparatus in order to form the various *N*–glycan structures required in glycoproteins.

The biosynthesis of the precursor lipid–linked oligosaccharide is generally referred to as the phosphodolichol pathway or dolichol cycle. It involves the sequential addition of sugars to the lipid carrier, dolicholphosphate. Dolichols are a family of long chain polyprenols, with one saturated isoprene unit ranging in size from C_{80} to C_{100}.

The dolichol cycle starts with the transfer of *N*–acetylglucosaminyl phosphate to phosphodolichol. The resulting diphosphate is then further glycosylated with a GlcNAc moiety, followed by the addition of five mannose residues. This series of reactions occurs in the endoplasmic reticulum (ER). Each of the transfer processes is catalyzed by five different membrane–bound glycosyltransferases. Interestingly, transfer of a further four mannose and three glucose moieties leading to the $Glc_3Man_9(GlcNAc)_2$–pyrophosporyl–dolichol, occurs from the respective phosphodolichol–activated carbohydrates and not from sugar diphosphonucleotides. Non–Leloir transferases are involved in these steps.

The dolichol anchors the growing oligosaccharide in the membrane of the endoplasmic reticulum. The exact topology of the dolicholphosphate cycle is rather complex and involves a flip–flop process in which the lipid–bound oligosaccharide is translocated from the cytosolic part of the ER to the luminal part. This is catalyzed by a 'flippase' as recently suggested by J. Helenius et al. (*Nature* 2002, *415*, 447). Then, the lipid–bound oligosaccharide is transferred to the growing, so–called nascending peptide chain. The latter reaction is catalyzed by an enzyme called oligosaccharyltransferase, that recognizes the amino acid asparagine as the acceptor in the tripeptide sequence Asn–X–Thr(Ser), where X can be any amino acid except a proline. However, not every asparagine in a consensus sequence Asn–X–Thr(Ser) is glycosylated, steric factors rather than just the correct sequence also determine the recognition process.

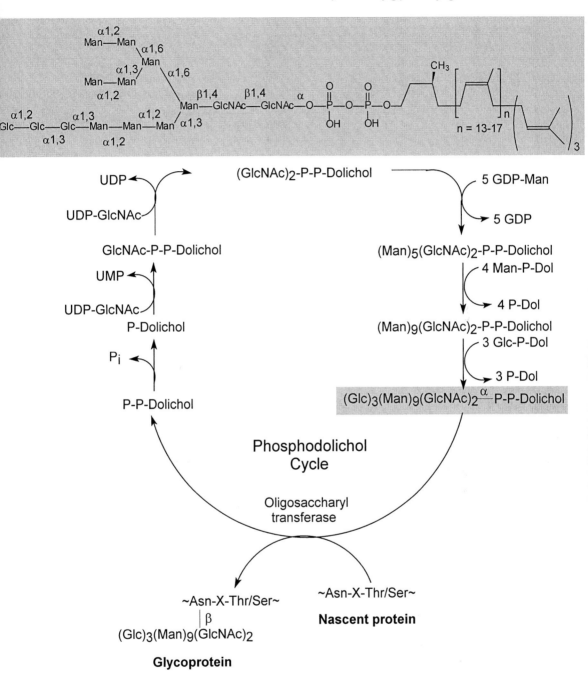

Schematic representation of the phosphodolichol cycle. The structure of the lipid–bound precursor saccharide, which is transferred onto the nascent protein, is highlighted in gray.

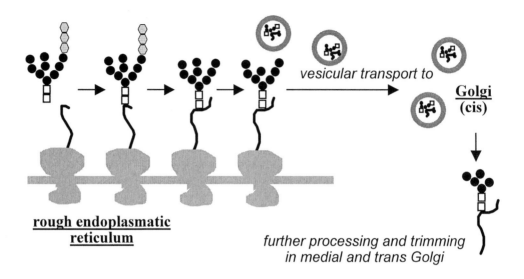

Schematic representation of the *N*–glycan biosynthesis. It takes place in cell organelles, the rough endoplasmic reticulum (RER) and the compartments (cis, medial, and trans) of the Golgi apparatus.

The size of the oligosaccharide part of a glycosylated protein is reduced by the action of specific glycosidases within the luminal part of the rough endoplasmic reticulum. The sum of these enzymatic steps is often referred to as 'trimming' and 'processing'. Processing of *N*–glycans at the ER is initiated by the removal of three glucose residues. Then, after one α–1→2–linked mannose residue has been removed, the glycoprotein is transported to the cis compartment of the Golgi apparatus by vesicular transport. The oligosaccharide is then processed to remove three more α–1→2–linked mannoses. The capped glycoprotein forms the precursor saccharide for the synthesis of high mannose–type oligosaccharides. On the other hand further mannose moieties can be split off and the truncated glycoprotein is further processed in the medial and trans compartments of the Golgi by various glycosyltransferases to form the different forms of complex and hybrid–type *N*–glycans. *N*–Glycans are then sorted and transported to their destinations in vesicles.

6.5   Biosynthesis of *O*–glycoproteins

The biosynthesis of *O*–glycans occurs differently from that of the *N*–glycans and does not follow a general scheme. Glycosylation of the complete peptide is carried out successively in the Golgi apparatus. It mostly starts with the transfer of an *N*–acetylgalactosamine residue. The glycan is then elongated and branched by other glycosyltransferases to form at least six core classes of different *O*–glycans (core I to core VI). L–Fucose, sialic acids or galactose residues are often attached as peripheral saccharides.

6.6 Biosynthesis of glycosphingolipids

While ceramide is synthesized in the cytosolic region of the endoplasmic reticulum, the attachment of the first monosaccharide, D–glucose, occurs on the cytosolic side of the Golgi membrane. The newly formed glucosylceramide then tunnels through the Golgi membrane and serves as a precursor in the formation of other, more complex glycolipids in the lumen of the Golgi apparatus.

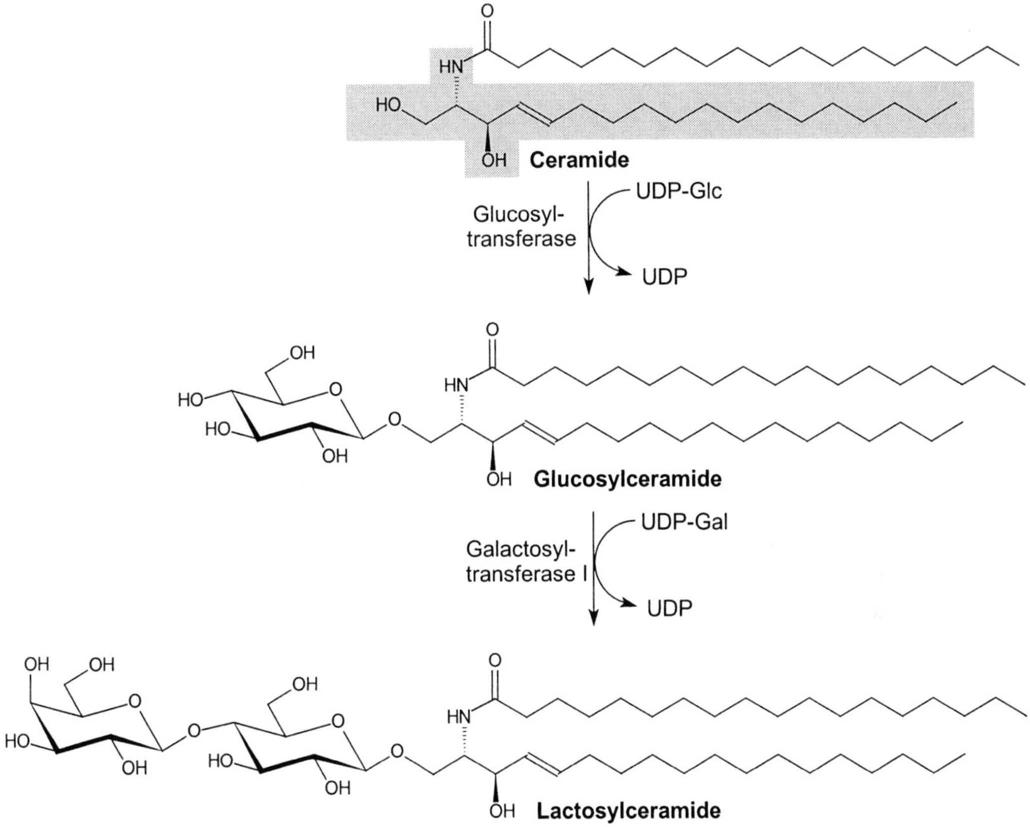

Biosynthesis of lactosylceramide. Biosynthesis of the ceramide (sphingosine part is highlighted in gray) starts from L–serine and palmitoyl–CoA and proceeds in the lumen of the endoplasmic reticulum. Glucosylceramide is formed at the cytosolic site of the Golgi apparatus. The glucosylceramide traverses the membrane into the lumen of the Golgi apparatus where further glycosylations takes place.

A relatively small set of enzymes catalyze the biosynthesis of the various glycosphingol-ipids, following parallel biosynthetic pathways originating from the gangliosides LacCer, GM3, GD3, and GT3. In the cis– and medial–compartments of the Golgi apparatus, mem-brane–bound, acceptor–specific glycosyltransferases are responsible for the production of precursors which are required for the synthesis of the different families of glycosphingolip-ids, such as lactosyl ceramide (LacCer), and the gangliosides GM3, GD3 und GT3.

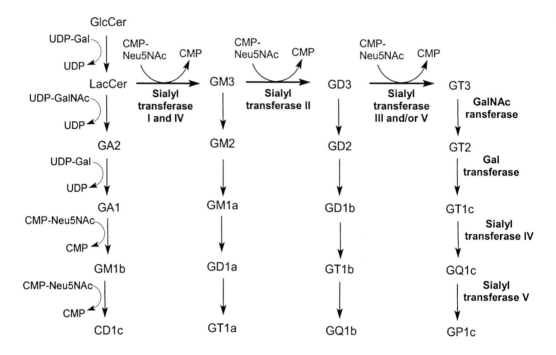

These are eventually converted to a number of different products by only a few glycosyl-transferases with broad acceptor specificity. In this way, the cell directs the synthesis of a large variety of glycosphingolipids with only a small set of enzymes. The products of these reactions are then transported to the plasma membrane surface by exocytotic vesicles.

7   Glycobiology

Glycobiology deals with the functions of sugars in biological recognition events. Thus, the topics of glycobiology are based on a large number of questions, such as:

(i)   What are the structures of oligosaccharides involved in molecular recognition?
(ii)   What are the structures of the molecules, which act as receptors for carbohydrate ligands?
(iii)   What are the molecular details of carbohydrate–receptor interactions?
(iv)   How is the formation of carbohydrate–receptor complexes translated into a biological effect?
(v)   How do carbohydrates influence the properties of proteins?
(vi)   Which functional assays can be used to elucidate the biological role of carbohydrates, and
(vii)   how can carbohydrates be used therapeutically?

It will require several decades to fully answer all these fundamental questions of glycobiology, however, many exciting details of the biological functions of carbohydrates have already been revealed and some of the highlights are covered in this chapter.

Nature uses the glycosylation of peptides to alter their properties as glycosylated proteins behave in a clearly different manner to that of their non–glycosylated counterparts in terms of solubility and stability, for instance. The enzyme ribonuclease (RNase) is an example of a protein, which exists *in vivo* both in non–glycosylated and glycosylated forms, the two molecules being named as RNase A and B. The oligosaccharide moiety of RNnase B shields a large section of the protein from interactions with other molecules. Thus, glycosylation contributes to the stability of RNase against proteases, but also decreases its activity because binding to the substrate (RNA) is hindered.

Many glycoproteins and glycolipids, are embedded in the outer membrane of cells (cf. section 6.2), representing major components of the outer surface of mammalian cells. The carbohydrate coat, surrounding a cell is characteristic of a particular species, the cell type, and its developmental status. Specific sets of *N*–glycans have been identified at distinct stages of cell differentation, and many alterations in cell surface oligosaccharides have been found to be associated with various pathological conditions including malignant transformation of cells and uncontrolled cell growth (cancer).

A common example of the importance of cell type–specific glycosylation is the existence of different blood groups, which is based on the different carbohydrate determinants present on the surfaces of blood cells.

7.1   Blood group specificities

The classical ABO blood group system provided the first example of single human poly-morphic characters that were not associated with inherited diseases. The understanding of the serological relationships in this system laid the basis for the safe transfusion of blood from one individual to another.

Table 7–1. The ABO blood group system.

Phenotype of red cells	Minimal determinant saccharide (blood group antigen)	Antibodies found in plasma	Glycosyltransferases expressed
A	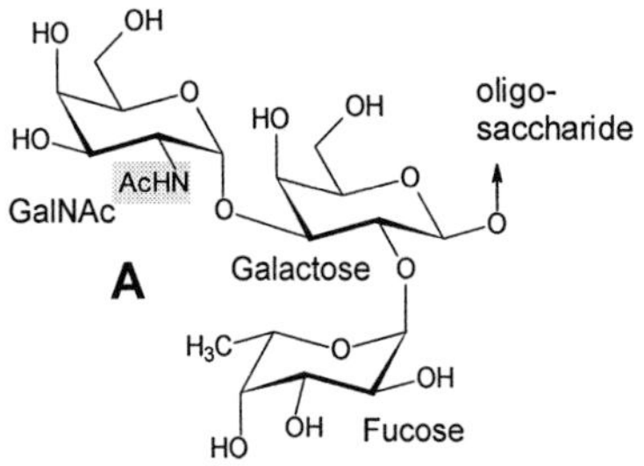 **A** — GalNAc, Galactose, Fucose, oligosaccharide (AcHN)	anti–B	α–1,3–GalNAc transferase
B	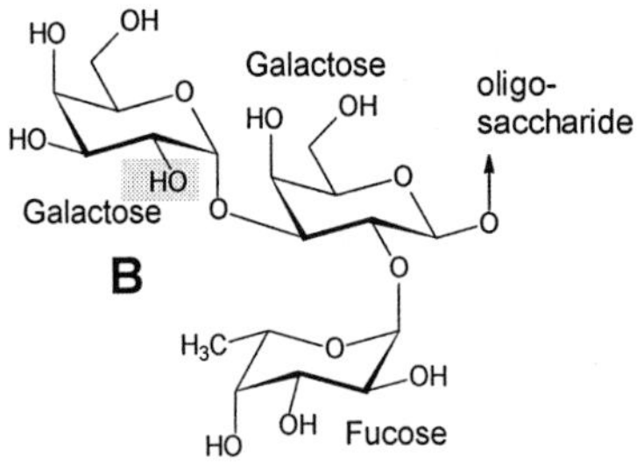 **B** — Galactose, Galactose, Fucose, oligosaccharide	anti–A	α–1,3–Gal transferase
AB	A and B	—	α–1,3–GalNAc transferase and α–1,3–Gal transferase
O	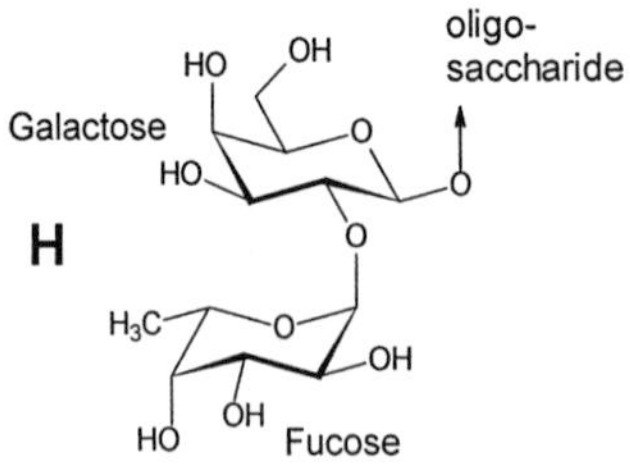 **H** — Galactose, Fucose, oligosaccharide	anti–A and anti–B	

The antigenic structures classified within the ABO blood–group system are oligosaccha-rides occurring as termini of glycoproteins, glycolipids and soluble oligosaccharides. These antigenic structures were termed 'blood–group antigens' because they were first discovered on the surface of erythrocytes, they are however, also found in other tissues and human secretions. The ABO classification is based on the presence or absence of two antigens, A and B, on the erythrocyte surface and two antibodies anti–A and anti–B which always occur in the plasma when the corresponding antigen is missing. Oligosaccharides, which are not further modified to A– or B–type structures are called phenotype O or H, and these were found to be recognized by an antibody which was called anti–H antibody.

The carbohydrate structures which are the basis of the ABH antigens are dependent on the inheritance and expression of genes encoding the glycosyltransferases, which determine their synthesis (Table 7–1). With rare exceptions, H is expressed on the cells of all group O individuals, having the terminal galactose moiety in the erythrocyte glycoprotein α–1,2–fucosylated. In persons belonging to group A, B, and AB further glycosylation with a GalNAc or galactose residue respectively, are catalyzed by the respective enzymes.

Glycoconjugate oligosaccharides, such as the blood group antigens, are exposed in a antenna–like manner to the cell exterior (cf. section 6.2). There, they are primed for interacting with other cells. They function as molecular addresses and by interaction with specific receptors, they can orchestrate a variety of biologically functions in intercellular communication such as cell proliferation, cell–cell adhesion and cell migration. Thus oligosaccharides contain biological information encoded in the form of their three–dimensional structures. Molecules, which can 'decode' the information stored in carbohydrates, are called lectins. These are proteins, which are specialized for the specific recognition of carbohydrate ligands resulting in the formation of non–covalent carbohydrate–protein (lectin) complexes.

7.2   Lectins

Many proteins occur in nature which can interact with carbohydrates non–covalently. These include carbohydrate–specific enzymes on the one hand, and antibodies on the other, which are formed as a reaction to the carbohydrate antigens encountered by the immune system. Lectins are a third class of carbohydrate–specific proteins. The name lectin is derived from the Latin *legere*, to pick or choose. Lectins specifically and reversibly bind monosaccharides, oligosaccharides, or partial structures of saccharides but are devoid of catalytic activity and also are not products of an immune response such as antibodies.

By the beginning of the century the first lectins had been discovered in plants and were later found to occur in almost all other organisms, ranging from viruses and bacteria to the animal kingdom. Each lectin molecule contains two or more carbohydrate–binding sites, which are known as carbohydrate recognition domains, abbreviated to CRD. Because lectins are di– or oligovalent with regard to the number of CRDs, they can cross–link cells by combining with carbohydrate moieties on two or more cell surfaces, thus forming cell precipitates, which are called agglutinates. The agglutination of erythrocytes, the highly glycosylated red blood cells, by lectins is called hemagglutination. Hemaggluatination is a major property of lectins and is often used for their detection and characterization.

Lectins represent a heterogeneous group of oligomeric proteins that vary widely in size, structure, molecular organization and especially in the their carbohydrate recognition domains. Nevertheless they can be classified according to similarities in their binding specificities and structural characteristics. Depending on the type of monosaccharides to which they exhibit the highest affinity, lectins have been classified into five groups, those specific (i) for mannose, (ii) for galactose and GalNAc, (iii) for GlcNAc, (iv) for L–fucose and (v) those specific for *N*–acetylneuraminic acid. Lectins are sometimes very specific, binding to only one kind of monosaccharide, for example, distinguishing between glucose and galactose or between GalNAc and GlcNAc. Therefore, lectins can also be used as tools to specifically 'fish' certain glycoproteins out of a glycoconjugate mixture. In other cases the specificity of a lectin for monosaccharides may be rather low, such as in the case of many lectins which are called 'mannose–specific' but also bind L–fucose. Moreover, lectins often do not selectively bind monosaccharide moieties, but interact with more complex oligosaccharide portions presented in glycoconjugates. Therefore, classification of lectins according to monosaccharide specificities as determined in *in vitro* experiments is a simplification of the natural specificities, which are often more complicated and not precisely elucidated.

Lectins occurring in animals consist of three subgroups, (i) the S–type lectins, (ii) the C–type lectins, and (iii) the P–type lectins. C–Type lectins are further classified into endocytic lectins, collectins, and selectins. Of the selectins, only three representatives are known so far, these are L–, E–, and P–selectin (cf. section 7.4).

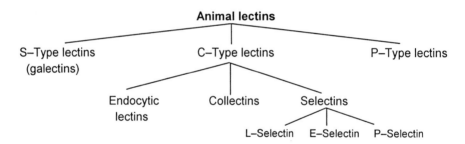

S–Type lectins are also called galectins. They are believed to participate in cell adhesion and are expressed in many different cell types and are found not only within the cyctoplasm and within the nucleus, but also outside the cell. They often require reducing groups such as thiols for binding. Galectins are specific for β–galactose moieties, such as those presented in lactose–terminated glycans, and many of them also recognize *N*–acetylglucosamine. Among the P–type lectins the most prominent example is the mannose–6–phosphate receptor, which serves for targeting of glycosylated lysosomal enzymes to their subcellular compartments. The C–type lectins are a large group of carbohydrate–binding proteins, which require Ca^{2+} in the binding process. They comprise endocytic lectins, collectins, and selectins.

The endocytic lectins are membrane–bound receptors with different carbohydrate specificities. One important example is the mammalian hepatic asialoglycoprotein receptor, discovered by Ashwell, which binds to galactose and GalNAc. It was shown to facilitate clearance from the circulation of those glycoproteins in which terminal sialic acid residues have been cleaved off and galactose or GalNAc residues are then terminally exposed (cf. section

7.3). Another prominent representative of endocytic lectins is the mannose–binding protein expressed on macrophages. Macrophages are important cells of the immune system which engulf and destroy infectious organisms after receptor binding. This type of defense, which does not depend on antibodies is called innate immunity. Collectins have similar functions as exemplified by the macrophage binding proteins. Another group of important endocytic lectins is the so–called mannose–binding proteins (MBPs), which are soluble lectins present in mammalian serum and liver. They bind to the oligomannosides of infectious microorganisms, causing activation of compliment without antibody participation, and subseuqent lysis of the pathogens.

Finally the selectins mediate the selective adhesion of circulating leucocytes to endothelial cells of blood vessels, a process which is a prerequisite for the removal of leucocytes from the circulation and for their migration into tissues where they are required (cf. section 7.4). The three types of selectins are known as L–selectin, P– and E–selectin.

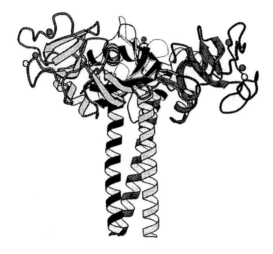

Ribbon diagram of the trimer MBP–A. The CRD of each monomer is located at the top of the figure, where small black circles indicate Ca^{2+} ions, which are present in the binding site and required for binding carbohydrate ligands.

Many lectins occur as oligomers such as MBP–A which occurs as a trimer. This permits binding of several carbohydrate ligands to the lectin, but also requires an appropriate spacing of ligands. Thus, many lectins do not bind monosaccharides at all, but bind only oligosaccharides in which the sugar moieties required for binding are assembled in the correct three–dimensional arrangement. Since oligosaccharides are mostly flexible molecules with considerable freedom of rotation around the glycosidic bonds, the affinities of lectins for these molecules may be influenced by their shape. Different lectins specific for the same oligosaccharide may recognize different regions on its surface. The selection of a specific three–dimensional oligosaccharide structure from the array of structures that exist with a dynamic molecule, is one of the most important features of lectin–sugar interactions which await elucidation.

7.3   Carbohydrate–protein interactions

Interactions of lectin receptors with carbohydrate ligands have been found to be of essential importance in many biological processes. Often the formation of carbohydrate–protein complexes form the initial step in a cascade of further receptor–ligand interactions, including protein–protein interactions, leading to a biological event such as extravasation of leucocytes or signal transduction. Lectins act as receptors in diverse biological processes, including clearance of glycoproteins from the circulatory system, control of intracellular traffic of glycoproteins, adhesion of microbes to their host cells, recruitment of leucocytes to inflammatory sites, and also in malignancy and metastasis. Table 7–2 lists a number of cell–cell recognition processes in which carbohydrate–protein interactions have been found to be essential.

Table 7–2. Carbohydrate–protein interactions in cell–cell communication.

Biological event	Carbohydrate ligands assembled on	Lectin receptors expressed on
Microbial infection	Host cells	Mircoorganisms
Immune response	Phagocytes, microorganisms	Microorganism, phagocytes, macrophages
Fertilization	Zona pellucida	Sperm
Leucocyte recruitment	Leucocytes, endothelial cells	Endothelial cells, leucocytes
Metastasis	Target organs, malignant cells	Malignant cells, target organs

Since lectin CRDs are typically flat binding pockets, carbohydrate ligand binding to the binding sites can only be superficial. For MBP, which mediates antibody–independent binding of pathogens containing a high concentration of mannose on their surfaces, the interactions with mannose have been determined from X–ray studies. The sugar is bound to a flat carbohydrate recognition domain via the chelation of Ca^{2+} ions to the 3– and 4–hydroxyl groups of mannose and the complex is further secured by hydrogen bonding to amino acid residues in the binding site. However, hydrogen bonds can also be formed between the CRD and water molecules in the absence of a sugar ligand, and consequently energy of ligand binding is low. The weakness of the carbohydrate–lectin interaction, accompanied by a relatively relaxed specificity is one of its most striking features.

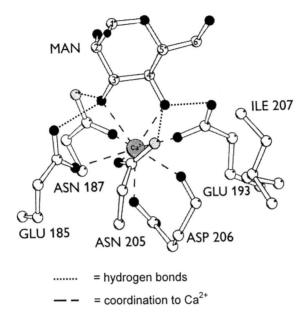

MAN

ILE 207

ASN 187

GLU 193

GLU 185

ASN 205   ASP 206

.... = hydrogen bonds

— — = coordination to Ca^{2+}

Binding of mannose in the carbohydrate recognition domain of rat MBP–A. Ca^{2+}, which is required for binding is coordinated to the oxygen atoms of surrounding amino acids as well as to the 3–OH and 4–OH groups of bound mannose. These two mannose hydroxyls also show hydroxyl bonds to the binding site. Thus mannose binding to the MBP–CRD is rather weak, as only 3–OH and 4–OH interact with atoms in the binding site.

The strength and specificity of carbohydrate–protein interactions are significantly improved by multivalent binding. The combination of the low binding affinities of each ligand leads to a higher binding avidity overall and thus biologically effective binding is established. (The term 'affinity' is used with regard to the strength of monovalent receptor–ligand binding, 'avidity' is used to describe the strength of association in a polyvalent interaction process). Multivalency of the receptor CRDs can be achieved by different means. Lectins may form oligomers as in the case of mannose binding protein (MBP) or asialoglycoprotein receptors from different animals; they may assemble more than one CRD on a single polypeptide chain, as in the macrophage mannose–receptor, which contains eight CRDs in one polypeptide. Moreover, lectin CRDs may be clustered in close proximity on the membrane by assembling monomeric lectins as is possibly the case with the selectins. On the other hand, the carbohydrates which serve as ligands for the lectins, are also available in multiple copies distributed in the glycocalyx of the cell surface. Moreover, the multivalency of carbohydrates in the multiantennary architecture of glycoconjugate oligosaccharides ensures that the geometry of the terminal sugars required for multiple binding to the receptors, is correct.

Multivalency is often an important principle in biology, and is particularly so in carbohydrate–protein interactions. There are a number of functional advantages with multivalent interactions such as fine–tuning of the biological response, giving rise to a spectrum of responses rather than just an 'on' or 'off' signal. Moreover, multivalency of receptor–ligand interactions can lead to greater contact between surfaces, resulting in conformational rearrangements of surface shapes which can then trigger signal transduction, for example.

In order to discriminate between endogeneous mammalian carbohydrates and those that are of microbial origin, lectins must have the ability to recognize a particular three–dimensional arrangement of a ligand. For example, while MPB binds bacterial oligomannosides, the innate immune response is not triggered by high mannose–type structures of mammalian origin.

The actual geometry of presentation of oligosaccharides to the receptor is important in order to trigger a biological response. In general, multiple interactions with multivalent targets are required for physiologically relevant binding. A structure presenting a single ligand will be bound and released, while multiple presentation of ligands will lead to more permanent binding.

A good example of the functional importance of multivalency in carbohydrate–protein interactions is the binding of desialylated glycoproteins to the asialoglycoprotein receptor. The asialoglycoprotein receptor in mammalian liver provided the first model of how an animal lectin might discriminate in a useful way between various glycoproteins. When erythrocytes are aging, they lose terminal sialic acids, thus exposing free galactose and *N*–acetylgalactosamine residues, these are recognized and bound by the asialoglycoprotein receptor, which is expressed on the surface of liver cells. Tight binding to the asialoglycoprotein receptor is only possible when multiple galactose residues are bound and in this way senescent erythrocytes are cleared from the blood circulation. Thus, the property of multivalent binding provides a mechanism by which aging erythrocytes can be discriminated from those, which are still functioning. The importance of multivalency for tight binding to the asialoglycoprotein receptor has been intensively investigated by Y. C. Lee using synthetic glycoclusters, with one, two or three galactose residues exposed. He found, that with the linear increase of sugar epitopes, binding to the receptor increased logarithmically. This effect he termed the 'cluster effect'. To obtain a cluster effect in the interaction of carbohydrate clusters and protein clusters, the sugar epitopes exposed in the (synthetic) glycoclusters must have a stereochemistry complementary to that of the receptor CRDs.

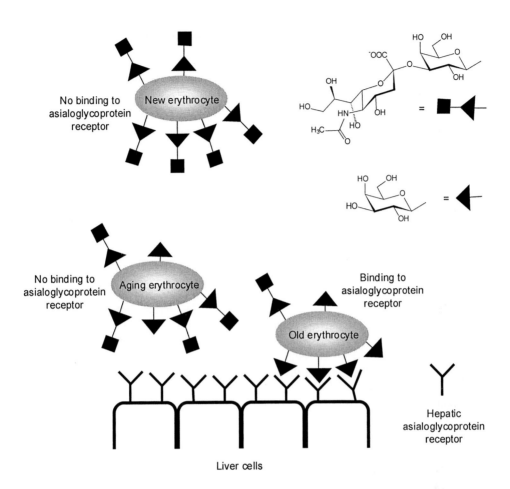

Diagram showing the removal of senescent erythrocytes from the circulation. When several terminal sialic acid residues have been cleaved off the cell surface glycans during normal cell aging, erythrocytes can be cleared from the blood circulation by multivalent interactions of the exposed galactose residues with a cluster of CRDs on the hepatic asialoglycoprotein receptor resulting in a tightly–bound complex.

7.4   Carbohydrate–selectin interactions in leucocyte trafficking

Clearly one of the most intensively investigated areas in glycobiology has been the role of carbohydrate–protein interactions in the early stages of inflammation. In cases of injury to the body or irritation of tissues, the immune system reacts by transporting the necessary cells to the sites where they are needed. The immune response is then accompanied by inflammation, the normal reaction of the body against invaders.

White blood cells, called leucocytes, are important cells of the immune system for repair of tissue damage and for defense against microbial invaders. They normally circulate rapidly through the body in the bloodstream. Their rate of circulation decreases at sites of injury however, to allow the cells to adhere to the endothelial cell layer which lines the blood vessels, and through which they eventually migrate. The early stage of leucocyte adhesion is mediated by carbohydrate–protein interactions, in which lectins, expressed both on leucocytes and on the endothelial cells bind to carbohydrate ligands located both on leucocytes as well as on endothelial cells. The lectins involved are called selectins in this case, as they mediate the selective adhesion of cells. Thus, carbohydrate–selectin interactions form the first step in a process which controls leucocyte trafficking to sites of inflammation. Three types of selectins are known, L–, P–, and E–selectin. L–selectin, sometimes also called the 'homing receptor', is found on all leucocytes and is also involved in the recirculation of lymphocytes, directing them specifically to peripheral lymph nodes ('homing'). P– and E–selectin are expressed on endothelial cells only when these cells are activated by inflammatory mediators.

The sialyl–Lewis–X (sLe^X) tetrasaccharide, a ligand for the selectins.

The carbohydrate ligands, which are bound by selectins, are located on both the leucocytes and endothelial cells. Multivalent carbohydrate–selectin interactions lead to a slowing down of the leucocytes, a process which has been termed 'rolling'. In a second step, firm adhesion is then facilitated by high affinity protein–protein interactions between the so–

called integrins on leucocytes and the intercellular adhesion molecule ICAM–1. This finally leads to flattening of the leucocytes and their migration through the endothelial layer, a process which is called vascular extravasation. Thus, carbohydrate–protein interactions are rapid but weak interactions which initiate leucocyte rolling, while the protein–protein interactions which occur later lead to firm adhesion and vascular extravasation.

It is not yet known which carbohydrate represents the optimal ligand for the selectins *in vivo*. However, a tetrasaccharide has been identified as the minimum oligosaccharide structure which is required for binding to the selectins. It is called sialyl–Lewis–X (sLe^X) and has the structure Neu5Acα2$\rightarrow$3Galβ1$\rightarrow$4(Fucα1$\rightarrow$3)GlcNAcβ–OH. SLe^X frequently occurs as the terminal end of numerous glycoconjugate oligosaccharides and because of its role in leucocyte adhesion it has become an important sugar and is used as lead structure for the development of mimetics which can effectively interfere in the early, carbohydrate–mediated stages of leucocyte adhesion in cases of acute inflammation (cf. section 7.6).

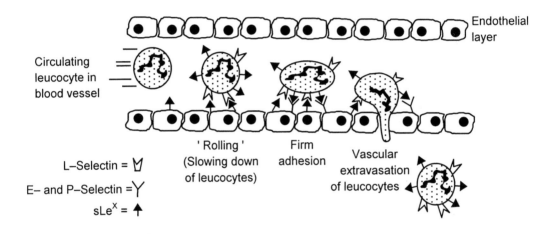

Schematic, simplified representation of leucocyte adhesion to endothelial cells. Only the interactions between selectins and carbohydrate (sLe^X) epitopes are shown, which are the important interactions in the early stages of leucocyte adhesion. To ensure the firm adhesion of leucocytes and vascular exravasation, strong protein–protein interactions are necessary, once rolling of the leucocytes is initiated.

7.5   Microbial adhesion

Carbohydrate–protein interactions not only have a function in physiological processes, but are also of importance in pathological processes, such as the adhesion of microbes, including viruses, bacteria, protozoa, or fungi to their host cells. Firm adhesion is important for microbes to succeed in infecting a cell, as they are otherwise swept away by the normal defense mechanisms of the body. Many microbial adhesion processes are dependent on carbohydrate–protein interactions. Thus viruses and bacteria, for example, use their own lectins to specifically interact with certain saccharides of the glycocalyx of a potential host cell.

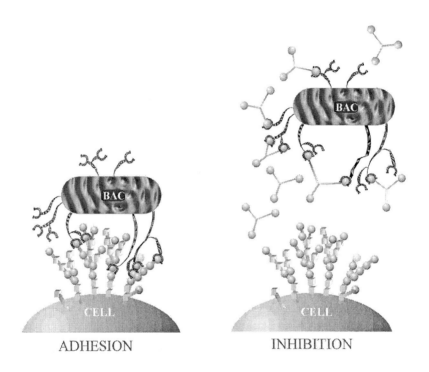

ADHESION   INHIBITION

Bacteria express a large number of lectins on so–called fimbriae and use these to adhere to the glycocalyx of a host cell in a multivalent interaction. This can be demonstrated by the agglutination of erythrocytes for example. Hemagglutination can be prevented by saccharides, for instance multivalent glycoconjugates, which bind to the bacterial lectins and thus inhibit bacterial adhesion.

Many bacteria possess long proteinaceous appendages which are called fimbriae or pili. They carry lectin domains which can be utilized by the bacteria for adhesion to host cells. Fimbriae are classified according to their carbohydrate specificities. The best characterized fimbriae are specific for α–mannosyl residues and are called type 1 fimbriae. They are common fimbriae of *Escherichia coli*. P fimbriae are also widely distributed and are specific for the disaccharide unit Gal–α(1,4)–Gal, called galabioside.

Bacteria often carry several hundred fimbriae on their surfaces, all with different carbohydrate specificities. Expression of fimbriae is often essential for the pathogenesis of bacteria and contributes to the virulence of particular bacterial strains.

Since adhesion is a prerequisite for infection, inhibiting the initial adhesion event would be a possible method of preventing infection. This could be achieved for example by using suitable multivalent glycomimetics as therapeutic anti–adhesives.

The attachment of viruses to the glycosylated surface of their host cells is also lectin–mediated. A well–documented example is *Influenza* virus. *Influenza* virus expresses two proteins on its surface, both of which are essential for the virus to succeed in infecting a cell. One protein is a lectin, which is called hemagglutinin and is specific for sialic acid residues. It mediates adhesion of the virus to the host cell surface. A second protein is a neuraminidase which is also essential for the virus in order to infect the cell. The approach to preventing influenza infections by blocking one of these two proteins has been intensively investigated. Hemagglutinin has been inhibited by multivalent glycoconjugates carrying sialic acid residues, this approach has not as yet been replicated *in vivo*. On the other hand a neuraminic acid–based inhibitor of the *Influenza* neuraminidase has been designed on the basis of the X–ray analysis of the crystallized enzyme. This inhibitor was successfully submitted to all clinical trials and has recently been introduced as a therapeutic against 'flu.

7.6   Glycomimetics

With the help of synthetic carbohydrates and glycoconjugates carbohydrate–protein interactions can be studied with regard to their structural and functional details. Moreover, carbohydrate–protein interactions can also be modulated or blocked in a therapeutic context.

One method of obtaining the necessary carbohydrate ligands for the manipulation of carbohydrate–protein interactions is their isolation from natural sources, which is, however, difficult because of the limited availability of suitable material and because of the microheterogeneity in which cell surface oligosaccharides occur. Synthesis of oligosacharides that occur naturally however, is a demanding and time–consuming task. Alternatively, sugar ligands can be designed as simplified carbohydrate derivatives, which are often called glycomimetics. The synthesis of glycomimetics has a number of potential advantages, (i) they are easier to prepare than the natural structures, (ii) their structural properties can be widely varied, (iii) they can be designed as non–biodegradable compounds and (iv) as low molecular weight derivatives. It cannot be guaranteed that glycomimetics will act as ligands for the receptors under investigation, and they might even display higher receptor affinities than their naturally–occurring counterparts.

The functional groups in sLe^X which are essential for selectin binding are highlighted and the face of the molecule which binds to the carbohydrate recognition domain of selectins is shaded in gray.

To achieve a rational design for glycomimetics, it is advantageous to know the structure of the receptor under investigation. If the receptor structures, as for most lectins, are not known, systematic derivatization of the ligands helps to identify the functional groups and stereochemical features that are essential for receptor binding. Considerable effort has been made to characterize the elements of the sLe^X tetrasaccharide which are crucial for binding to the selectins.

It was found, that particularly the fucose moiety and the negative charge provided by neuraminic acid were essential for binding to the selectins, while the GlcNAc residue for example, is not necessarily required. An enormous variety of sLe^X glycomimetics have been designed on basis of these findings and have been tested for their ability to bind to the various selectins. Ideally sLe^X mimetics would interfere in the early, carbohydrate–mediated stages of leucocyte recruitment at low concentrations. These molecules could then be used as anti–inflammatory agents in disease states which are accompanied by chronical or acute inflammation, where overrecruitment of leucocytes causes damage to normal cells and would be important in the treatment of stroke, asthma and arthritis, for example.

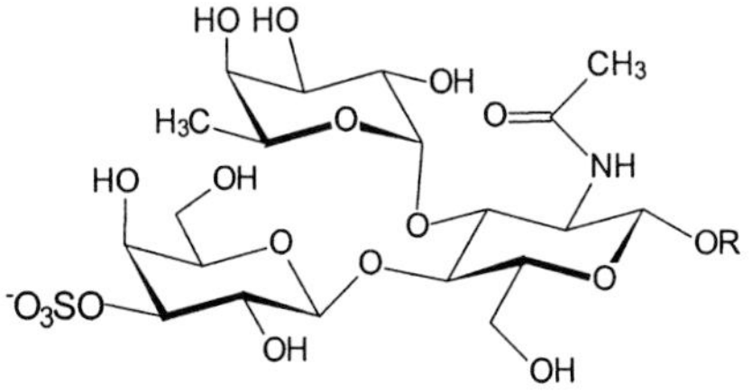

3'-sulfo-sialyl-Le^X

One of the first steps en route to simplified sLe^X derivatives is replacement of the neuraminic acid with a sulfate group, resulting in the preparation of 3'–sulfo–Le^X. 'Stripping' of the sLe^X lead structure went so far however, that the resulting sLe^X mimetics could hardly be regarded as oligosaccharides anymore. Some of these compounds were quite successful as inhibitors of leucocyte adhesion. While sLe^X shows IC$_{50}$ values between 1 millimolar and a few millimolar depending on which of the three selectins is used as the receptor, the amide shown below, which is based on a 6–*O*–hexadodecyl–modified *C*–galactoside was shown to have significantly lower IC$_{50}$ values (IC$_{50}$ values are ligand concentrations at which receptor binding is inhibitited to an extent of 50%).

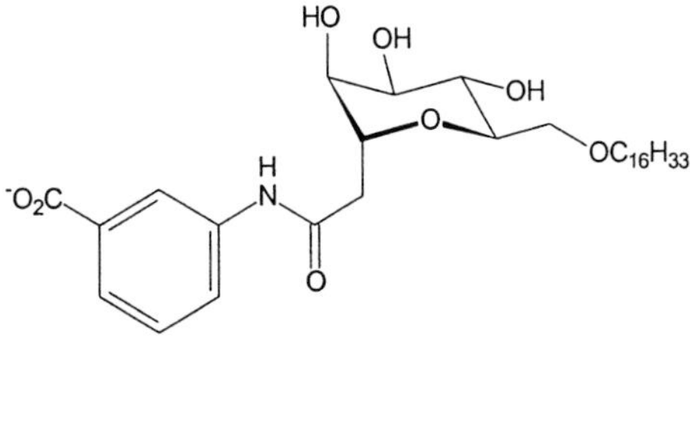

When sLe^X glycomimetics are tested as antiinflammatory agents, the assay used for the determination of selectin affinities is of crucial importance, as *in vivo* activities usually differ dramatically from results obtained in various *in vitro* assays.

Another important approach to the design of synthetic glycomimetics and glycoconjugates is the preparation of multivalent glycoconjugates which mimic the multiantennary nature of the oligosaccharides found in naturally–occuring glycoconjugates. Multivalent neoglycoconjugates have been designed with respect to the multivalency principle which is of functional importance in most carbohydrate–protein interactions.

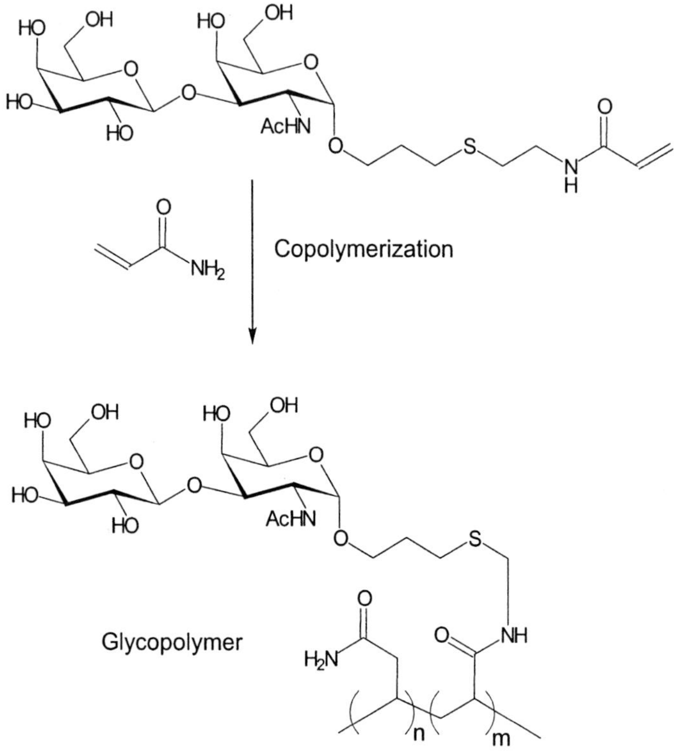

Example of a glycopolymer synthesis by the copolymerization of a functionalized Galβ(1,3)-GalNAc (the so–called T antigen disaccharide) with acrylamide.

They have been synthesized in many different variations, ranging from polydisperse to structurally distinct, so–called monodisperse derivatives. One of the most challenging problems when mimicking oligoantennary oligosaccharides is the choice of an appropriate scaffold to provide the appropriate three–dimensional arrangement of the assembled sugar ligands. The basic question regarding the ideal scaffold for sugar clustering is far from being answered conclusively.

Polymers, peptides, calixarens and dendrimers have been used as core molecules for the synthesis of multivalent glycoconjugates. Polydisperse molecules have been designed as 'glycopolymers' and the smaller 'glycotelomers', monodisperse multivalent glycoconjugates have been synthesized as so–called 'glycoclusters' and 'glycodendrimers', respectively.

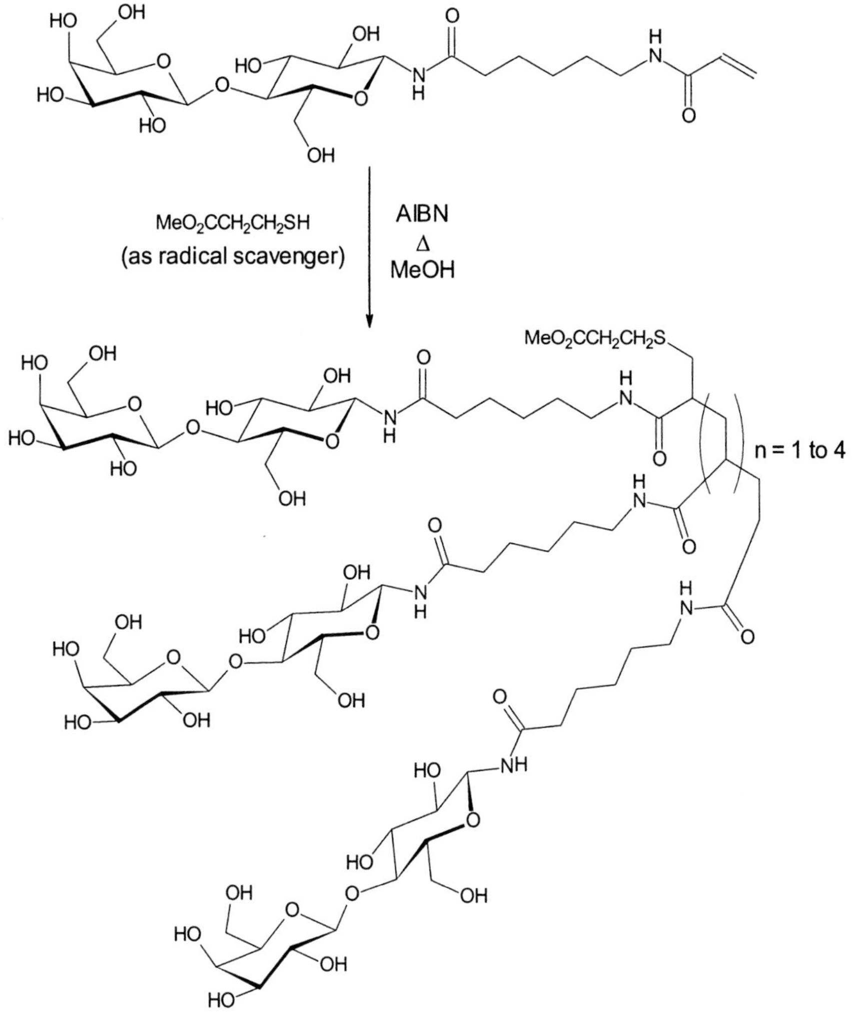

Short glycopolymers, called telomers can be obtained by radical homopolymerization of acrylamide–modified saccharide derivatives and addition of thiols as radical scavengers as shown here with a lactose derivative. By varying the amount of radical scavenger the number of repeating units can be controlled.

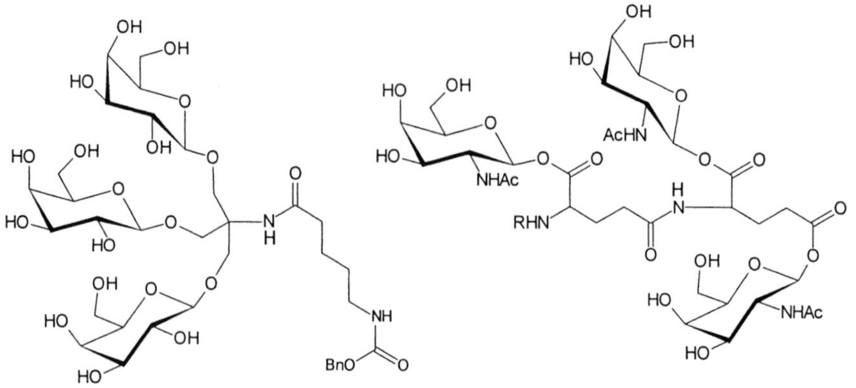

Examples of trivalent glycoclusters presenting galactose and GalNAc moieties, respectively with different stereochemistries. These derivatives were shown to bind to asialoglyoprotein receptors in nanomolar concentrations, representing an impressive example of the 'glycoside cluster effect'.

In all these examples the complex inner region of glycan structures has been replaced by a simpler branched, oligofunctional core molecule, which can be a polymer or telomer, an oligofunctional core molecule such as tris(hydroxymethyl)methylamine (TRIS), a peptide or a peptoid, respectively. In a glycopeptoid the amide nitrogen atoms of an oligoglycine molecule are functionalized with carbohydrates.

Peptide

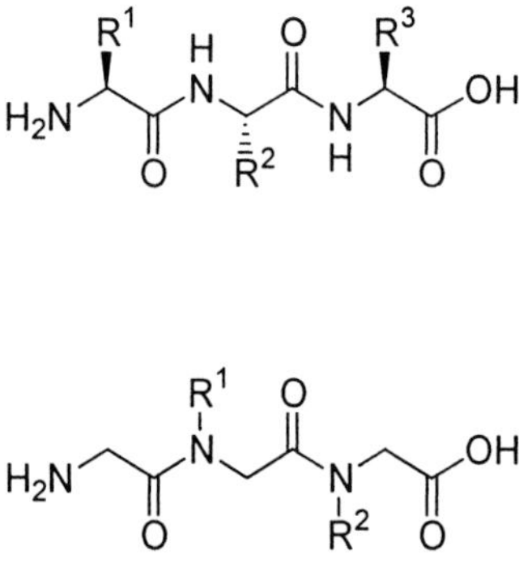

Peptoid

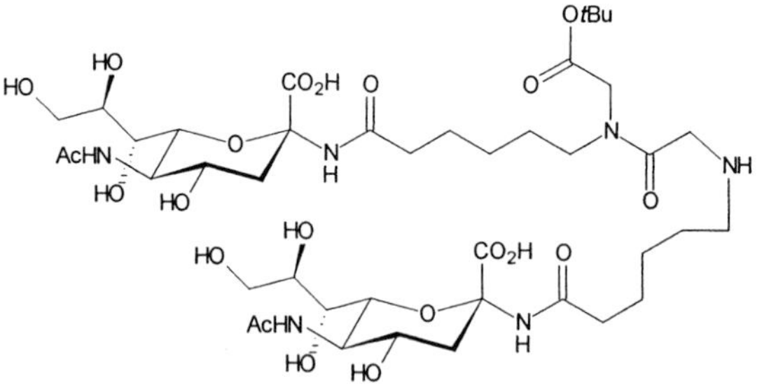

Peptoids differ from peptides, in that their peptide backbone consists of the amino acid glycine only, making it possible to assemble substituents (R) at the nitrogen atoms of the peptide bonds. Sialidated glycopeptoids have been synthesized by peptide coupling of building blocks as shown, and the sialoclusters obtained were tested as inhibitors of the adhesion of *Influenza* virus to its host cells.

Also monodisperse hyperbranched molecules, called dendrimers, have been used for car-bohydrate clustering leading to so–called glycodendrimers. As dendrimers can be grown generationwise, glycodendrimers with defined valency and spacing can be prepared easily. For attachment of the saccharide portions to the core molecules, linkages other than glyco-sidic bonds have been used. For example, using amino–terminated polyamidoamine (PAMAM) dendrimers as core molecules, reaction with isothiocyanato–functionalized carbohydrates, such as glycosyl isothiocyanates, leads to thiourea–bridged glycodendrimers in a reaction which requires no stereocontrol at the anomeric center.

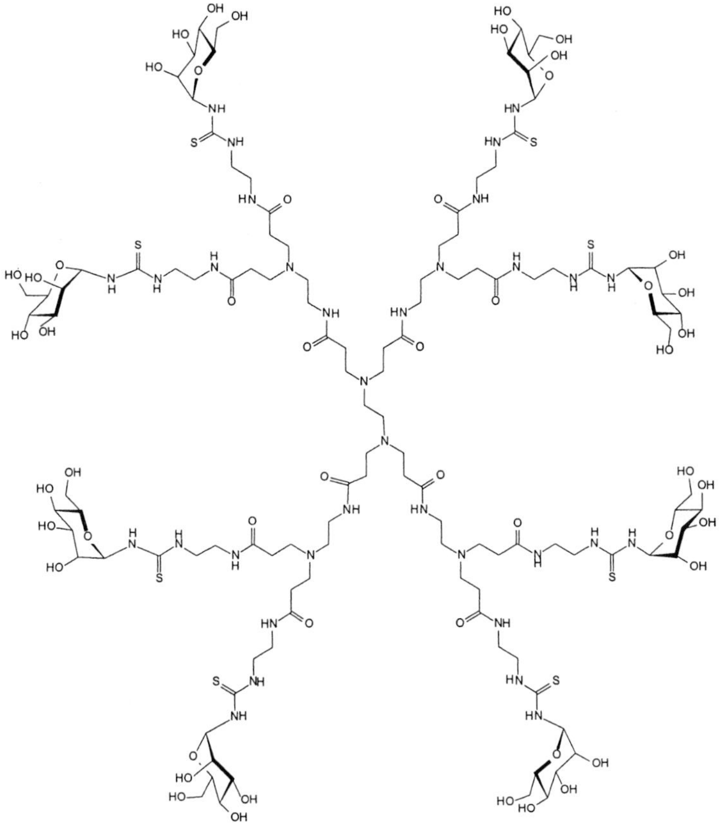

Example of a small glycodendrimer, where a second generation polyamidoamine (PAMAM) dendrimer was coated with α–mannosyl moieties by formation of thiourea bridges. Compounds such as that shown can inhibit mannose–specific adhesion of *Escherichia coli* at relatively low concentrations.

7.7   Outlook

Today we are aware of the fact that carbohydrates have more biological functions than just serving in energy storage and as architectural material. Through the work of glyco-chemists and glycobiologists we have received insight into quite a variety of processes in cell communication which are orchestrated by sugars, and we feel the potential and the excitement in what we have learned. However, one may ask, how much of the cell biology 'music' have we actually really heard and understood so far?

It appears to be proven that complex carbohydrates on cell surfaces resemble ligands for recognition by specialized proteins, the lectins (and selectins), to form noncovalent ligand-receptor complexes. Certain structural cut-outs of the glycocalyx carbohydrates are selected as epitopes to bind to the lectin CRDs (carbohydrate recognition domains) more or less specifically. However, by reinvestigating the *in vivo* situation, it becomes clear that the buzz of activity on the cell surface is still far from being understood. The cell surface environment is much more complicated and puzzling than, e.g., ligand binding to a discrete pharmacologically relevant receptor, as exploited in the classical fields of medicinal chemistry.

If the formation of ligand-receptor complexes to trigger a certain biological response, such as signal transduction, for example, is really the central task that the glycocalyx carbohydrates are meant to take over, how are the enormous redundance, flexibility, structural dynamic and microheterogeniety understood, which are such obvious characteristics of the sugar environment of cell surfaces? This does not fit with just receptor binding; at least, the conditions for specific receptor binding are unfavorable on the cell surface.

Therefore, it is possibly again time for glycobiologists to raise provocative questions and suggestions as they did earlier, when carbohydrates were still doomed to be not much more than table sugar. Thus, one may ask, what could carbohydrates do other than just providing ligands for receptor binding? This enormously large array of complex structures covering a cell could possibly form a blueprint of the respective cell life, reflecting its status, its history, its 'experiences', and thus storing and remembering some of the chemistry as well as physics which may befall a cell during its lifetime. This may be a central part of what is to be discovered next in cell biology. Water and water clusters might very well be an important part of this game.

One might call these lines the blueprint hypothesis or purely speculative. Nevertheless, it is my aim to encourage my readers and students to think beyond what has already been considered possible. Carbohydrates, even nowadays, form a typical field for discovering new knowledge.

8 Purification and analysis of carbohydrates

Many carbohydrate derivatives are syrups rather than solids and therefore, they cannot be purified by recrystallization procedures. This problem severely hampered early carbohydrate chemistry, and the situation only changed with the introduction of column chromatography. Column chromatography facilitates the separation of a mixture of syrupy derivatives and thus has become an indispensable tool for carbohydrate chemists.

The structural analysis of sugars, on the other hand, was revolutionized by NMR spectroscopy. The improvement of NMR instruments and NMR methods during the second half of the twentieth century has contributed immensely to the development of carbohydrate chemistry. The aim of this chapter is mainly to show how this set of essential methods is used in carbohydrate chemistry.

8.1 Chromatography and polarimetry

Thin layer and column chromatography

The principle of separating sugars of similar structure both by thin layer as well as by column chromatography is based on their different distribution between a mobile phase and a stationary phase. A suitable mixture of (mostly) organic solvents is selected as the mobile phase for each individual separation problem The most widely used stationary phase is silica gel. It is suitable for a wide variety of compounds of varying hydrophobicity. For highly polar derivatives, modified silica gel is used for chromatography, in which the Si–OH groups are etherified with alkyl chains. This material is called reverse phase silica gel and known as 'RP–8' or 'RP–18', depending on the length of the hydrocarbon chain used for etherification.

Thin layer chromatography (TLC) is used to monitor the course of a reaction and to qualitatively and also semi–quantitatively estimate the result of a particular conversion by comparison of the reaction products with authentic samples on a TLC plate. Thus, thin layer chromatography is a tool used on a daily basis in the laboratories of carbohydrate chemists.

TLC is mostly carried out on thin glass or aluminum plates, which are coated with a layer of the stationary phase. A wide choice of adsorbents is commercially available in addition to finely powdered silica gel, including reversed phase silica or other silica gel modifications, alox, cellulose and cellulose derivatives, and celite.

TLC is carried out as an ascending chromatography on plates approximately 10 cm high. A pencil line is drawn across the bottom of the thin layer plate and a suitably diluted sample of the compound or compound mixture is spotted along the line, in some cases together with standard reference samples. The spots are dried and the chromatogram is placed into a cylindric flask, the bottom of which has been covered with the mobile phase, so that its ends are not touching the glass wall and the starting line is above the solvent surface. The chamber is closed and the solvent ascends the TLC plate by capillary force. Before the solvent front reaches the top, the TLC plate is removed from the chamber, the solvent front is then marked with a pencil line and the plate is dried carefully and submitted to one or several suitable detection methods. The most common reagent used for the detection of carbohydrates is 10–20% sulfuric acid in ethanol. The TLC plate is either dipped into such a solution or sprayed with it, followed by heating. Many derivatives with functional groups that absorb UV–light are visualized on the chromatogram using a UV–lamp.

The rate of movement of a particular compound in thin layer chromatography is described by an R_f value. This is the ratio of the distance the compound migrates to the distance the solvent moves from the starting point, i.e. the solvent front.

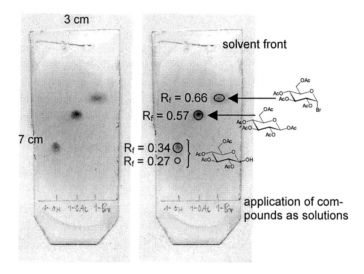

An example of TLC in which the reducing sugar, 2,3,4,6–tetra–O–acetyl–glucose, the corresponding pentaacetate and 2,3,4,6–tetra–O–acetyl–α–D–glucopyranosyl bromide have been distinguished. The TLC plate is labelled on the copy given on the right, indicating R_f values and structures for every spot. In the case of the 1–OH–free glucose derivative a small amount of the second anomer can be seen as shadow below the main spot. This TLC has been performed on silica gel, with ethyl acetate–toluene (1:1) as the mobile phase. Detection of the compounds was achieved by dipping the plate into 10% ethanolic H_2SO_4 followed by heating.

TLC can be modified as a two–dimensional method, in which the sample is spotted in the lower left–hand corner and chromatographed in the first dimension with solvent A. The chromatogram is then removed from the solvent, dried, turned through 90°, and chromatographed in the second dimension with solvent B, giving a two–dimensional separation of the compound mixture.

The eluant, which is best suited for the separation of a mixture by column chromatography is a solvent mixture which produces R_f values around 0.3 in TLC. The column used for the purification is filled with the same material, as used for TLC, normally silica gel, and equilibrated with the mobile phase. A concentrated solution of the product mixture in the mobile phase is then applied to the column and eluted with the solvent. Fractions of suitable size are collected in test tubes, analyzed by TLC, and pure fractions containing one type of derivative are combined and the solvent is removed *in vacuo*.

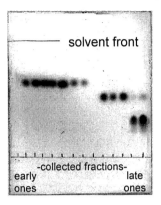

An example of TLC, used to check the result of purification by column chromatography on silica gel. Visualization of the spots was achieved by 10% ethanolic H_2SO_4 and heating. Three carbohydrate derivatives were separated. The compound with the highest R_f value elutes first, the one with the smallest R_f value is the last to be collected. To improve the performance of column chromatography, solvent gradients may be used, in which the polarity of the eluant is slowly increased.

The success of thin layer and column chromatography is decisively dependent on experience and the right choice of solvent mixtures used as the mobile phase. The relation between the stationary and mobile phases, polarity of solvent and products to be separated and their R_f values are compared in Table 8–1.

Table 8–1. Comparison of normal and reversed phase silica gel for chromatography.

Stationary phase	Compound	Mobile phase	R_f value
Silica gel Polar Si–OH	Non–polar (hydrophobic) compounds have higher R_f values than more polar compounds	Solvent mixtures of higher polarity increase R_f values	Higher for non–polar compounds than for polar compounds Higher with more polar solvent mixtures
Reversed phase silica gel Non–polar Si–OR	Polar (hydrophilic) compounds have higher R_f values than non–polar compounds	Solvent mixtures of lower polarity are used and increase R_f values	Higher for more polar compounds than for non–polar compounds Higher with non–polar solvent mitures

The performance of chromatographic purifications has been enormously improved by the introduction of HPLC (high pressure liquid chromatography). This technique, however, requires expensive instrumentation and extensive operator training. The chromatographic method most widely used in synthetic carbohydrate chemistry is so–called flash chromatography which gives good results and in which fine silica gel is used in columns to which a maximum pressure of 1 bar is applied. This improves the purification and accelerates the separation procedure significantly.

Very large molecules cannot be separated on silica gel. They are normally purified on gel columns, which separate compounds according to their size. This method is called gel permeation chromatography (GPC) and is carried out with a variety of gel materials which can tolerate the use of organic solvents as eluants on the one hand, and the use of water and aqueous buffer solutions on the other.

Chiroptical methods

Polarimetry was of fundamental importance for the analysis of sugars in the early days of carbohydrate chemistry (cf. chapter 2). Rotation values are still reported today in the literature for new carbohydrate derivatives however, they are now of minimal importance in the characterization of sugars.

Sugars, like all asymmetric compounds cause rotation in the plane of polarization of linear polarized light because the two circular polarized components travel at different speeds through it or a solution of it. The optical activity of an asymmetric sample in solution is observed in a polarimeter and measured as a rotation value, α. It is reported as specific rotation $[\alpha]$ obtained at a certain temperature and wavelength λ. The observed α and the specific rotation $[\alpha]$ are related in the manner shown in the formula

$$[\alpha]_{\lambda}^{\circ C} = \frac{100 \cdot \alpha_{\lambda}^{\circ C}}{l \cdot c}$$

where l is the length of the cell in decimeters (cells are often 1 dm in length) and c the concentration of the compound given in g per 100 ml.

Consequently, the unit for specific rotation values is (10^{-1} deg cm^2 g^{-1}) but has traditionally been reported, incorrectly, simply in degrees. The observed rotation value α depends on the concentration of the sample, the temperature and wavelength used, the latter usually being the wavelength of the sodium–D (double) line, 589 nm. Specific rotation values are, therefore, reported together with these specifications, for example:

$$[\alpha]_{D}^{26} = +160.5° \ (c = 1.09 \text{ in MeOH}).$$

The specific rotation value has the same sign as the observed α of a particular compound. Optically active compounds, which turn the plane of linear polarized light to the right, are assigned a positive sign for the observed rotation value. The enantiomer of this compound gives the same rotation value in the polarimeter but with the opposite sign. Rotation values of compounds of different molecular mass, can be calculated using an equation which relates specific rotation values to molar rotation values: $[M] = M \ [\alpha] / 100$.

Polarimetry is the traditional method of establishing the non–racemic nature of a sample of a chiral compound. It still finds a use today in the estimation of the anomeric purity of glycosides for example, because α– and β–glycosides normally show very different rotation values. Polarimetry is however, no longer used for the characterization of carbohydrate derivatives. Other chiroptical techniques, such as ORD (optical rotation dispersion) and CD (circular dichroism) provide more information than polarimetry.

For the structural elucidation of carbohydrate derivatives, NMR spectroscopy is the method of choice and has become an indispensable tool for the carbohydrate chemist. It is even more widely used than in some other areas of organic chemistry.

8.2   NMR and mass spectroscopy

NMR spectroscopy

NMR spectroscopy of simpler carbohydrate derivatives is primarily concerned with the hydrogen atoms of the sugar ring(s). There are three fundamental sets of information which can be obtained from the ^{1}H NMR spectrum of organic compounds: (i) the chemical shift of the signals, δ; (ii) their integration; and (iii) the coupling constants J deduced from the detected multiplets.

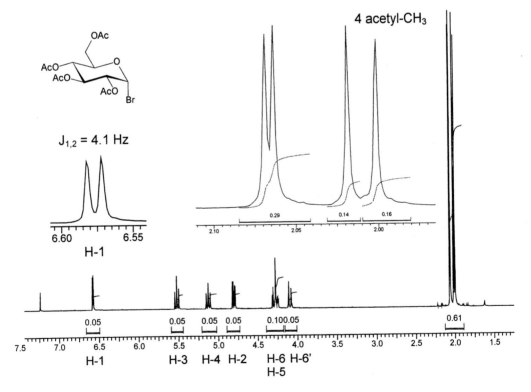

^{1}H NMR spectrum of 2,3,4,6–tetra–O–acetyl–α–D–glucopyranosyl bromide, recorded in CDCl$_3$ at 400 MHz. The doublet for H–1 and the singulets for the acetyl groups are expanded.

The chemical shift of a particular ring hydrogen atom is strongly dependent on whether it is an anomeric or a non–anomeric hydrogen and, furthermore, on the substitution pattern of the adjacent groups, especially of the functional group connected to the same carbon atom as the resonating hydrogen. The ^{1}H NMR spectrum of 2,3,4,6–tetra–*O*–acetyl–α–D–glucosyl bromide shown above, is typical for a monosaccharide derivative. The anomeric proton is characterized by its extreme down–field shift and appears as doublet, coupling with H–2. It is always the most easily assigned hydrogen atom of the carbohydrate ring as it is bound to the ring oxygen atom and, in the case of glycosyl bromides, is also bound to an electronegative bromine substituent. Both substituents withdraw electrons, thus deshielding the anomeric hydrogen and causing its signal to appear in most cases at lower field than all other ring hydrogens. For the other ring protons, typical chemical shift differences in the ^{1}H NMR spectrum are also observed as a consequence of their different locations in the sugar ring. In peracetylated sugars H–2, H–3, and H–4 show very similar chemical shifts, while both H–6 protons, often referred to as H–6 and H–6' (or H–6a and H–6b, respectively) are shifted up–field; H–5 shows the smallest ppm–value of all ring protons. The signals for H-2, H–3, H–4, H–6, and H–6' are dd–patterns as each of them couples with two neighboring protons, e.g. H–2 couples with H–1 and H–3. The protons H–2, H–3, and H–4 of a pyranose ring appear as pseudo–triplets instead of regular dd–systems, when both coupling constants are of almost the same size. H–5 gives a ddd–system, coupling to H–4, H–6, and H-6'. The acetyl–CH$_3$ protons of acetylated derivatives appear as singlets at around 2 ppm and are useful reporter groups in a molecule. A similarly useful landmark signal is the singulet obtained for the aglycone group of methyl glycosides, resonating around 3.5 ppm. Based on the different chemical shifts and observed coupling patterns of sugar protons the hydrogen skeleton of a particular compound can be reconstructed and normally, given the experimental details of the chemical reaction, the structure and conformation of the reaction product can be fully determined by NMR analysis.

Comparing the ^{1}H NMR spectra of α– and β–penta–*O*–acetyl–glucose, differences in the chemical shift of H–1 and the size of the coupling constant $J_{1,2}$ are significant. As a rule, an equatorial hydrogen substituent will give rise to a signal at lower field than an axial hydrogen when they are in a similar environment and the molecules are in the same conformation. Consequently, the H–1 of a α–D–configured sugar derivative has higher ppm–values than

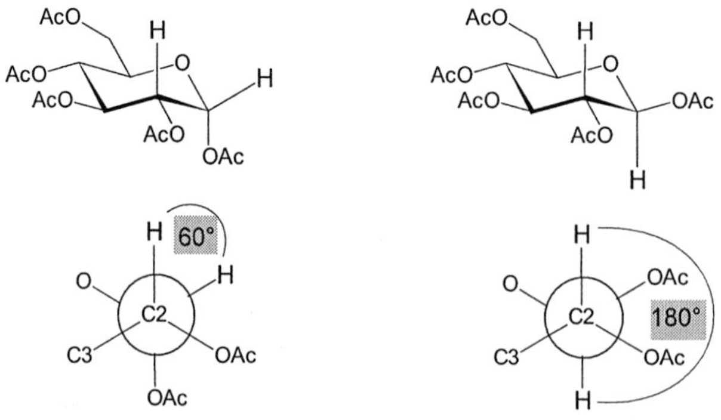

1,2,3,4,6-Tetra-O-acetyl-
α -glucopyranose
$J_{1,2}$ = 3.5 Hz

1,2,3,4,6-Tetra-O-acetyl-
β -glucopyranose
$J_{1,2}$ = 8.1 Hz

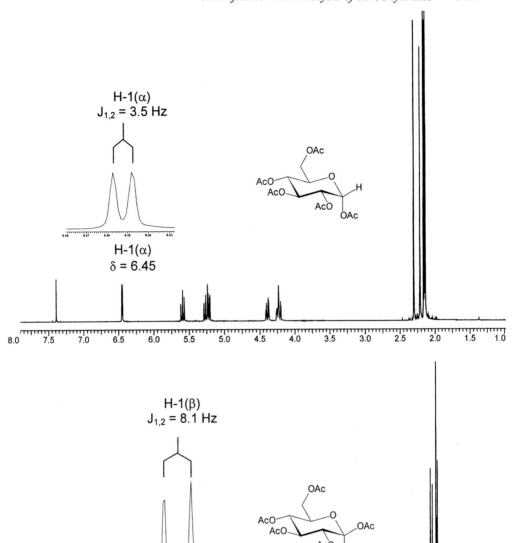

^{1}H NMR (400 MHz, CDCl$_3$) spectra of α– and β–penta–O–acetyl–glucose. The doublet for H-1 is expanded in each case and its chemical shift and coupling constant $^3J_{1,2}$ are indicated.

the H–1 of the β–D–configured analog, with chemical shift differences around 0.5 ppm. The difference in $J_{1,2}$ is caused by different dihedral angles ϕ between H–1 and H–2. The dihedral angle is 60° in the α– and 180° in the β–analog, respectively.

The dependence of the size of coupling constants ^3J (the superscript '3' indicating coupling over three bonds) on the dihedral angle ϕ of vicinal hydrogen atoms was first recognized by Karplus. According to Karplus, *trans*–diaxial hydrogens, with $\phi= 180°$ have a large coupling constant J, whereas axial–equatorial and equatorial–equatorial orientations, both having $\phi = 60°$, have much smaller coupling constants. Karplus identified the relationship between J and ϕ by showing that the values for J are approximately proportionate to ϕ and can be plotted against $\cos^2\phi$ to produce a characteristic curve. ^3J couplings occurring in regular six–membered rings in chair conformation fit the Karplus curve well. However, the size of coupling constants is also dependent upon the presence of strongly electron–withdrawing groups. With increasing electronegativity of a substituent a decrease in the value of the coupling constants is observed.

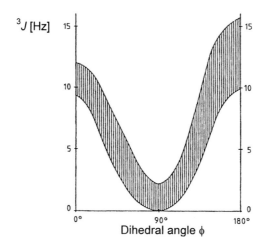

The graph shows the Karplus relationship between size of coupling constants ^3J and the dihedral angle ϕ of vicinal hydrogen atoms. In six–membered rings with chair conformation the size of ^3J is found somewhere within the area hatched in gray.

The Karplus relationship is of major importance for the interpretation of NMR spectra of carbohydrates as the conformation of the sugar and the relative configuration of the stereogenic centers in the pyranose ring can be deduced from the values of ^3J. Karplus' relationship is an especially valuable tool for the determination of the anomeric configuration of a sugar derivative. Moreover, many synthetic derivatizations of the sugar ring include a change of the configuration at one or several stereogenic centers and in these cases the Karplus relationship provides an excellent method of determining the stereochemical outcome of the reaction as shown by the ^3J proton–proton coupling constants.

In mannosides, carrying an axial 2–OH group, the size of $^3J_{1,2}$ is not indicative of the anomeric configuration. The coupling constant between H–1 and H–2 is small for both, α– and β–mannosides as their dihedral angle is 60° in any case. To prove the anomeric conguration of mannosides by NMR, a different criterion has to be employed, therefore. This is the size of the hetero–coupling constant $^1J_{C-1,H-1}$. For glycosides with different anomeric configurations a difference of approximately 10 Hz in the size of $^1J_{C-1,H-1}$ is observed. The anomer that has the equatorially disposed hydrogen at C–1 has the larger coupling constant. The $^1J_{C-1,H-1}$ value in such pyranosides is in general ~170 Hz, whereas those having an axially oriented hydrogen at C–1 have $^1J_{C-1,H-1}$ couplings of ~160 Hz. For mannosides both of these values are often found to be ~5 Hz smaller. $^1J_{C-1,H-1}$ values are determined in so–called gated–decoupling experiments, in which all C–H hetero coupling constants can be observed.

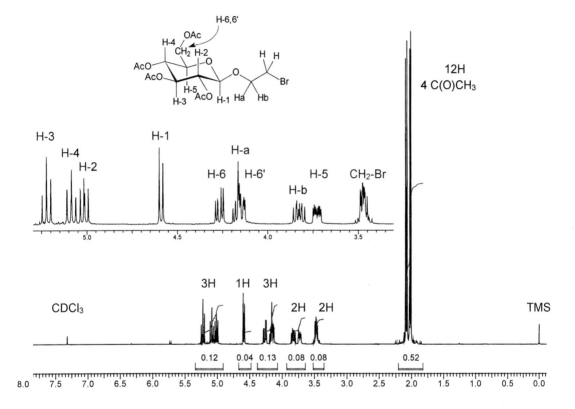

^{1}H NMR (400 MHz, CDCl$_3$) spectrum of (2–bromoethyl) 2,3,4,6–tetra–O–acetyl–β–D–gluco–pyranoside. The region of the ring protons is expanded. All peaks in the spectrum can be fully assigned on basis of the recorded chemical shifts and coupling constants. These data, which are obtained by NMR spectroscopy are always reported in the literature, when novel derivatives are described, and are shown in the following manner: **(2–Bromoethyl) 2,3,4,6–tetra–O–acetyl–β–D–gluco–pyranoside** (for synthesis cf. chapter 4): ^{1}H NMR (400 MHz, CDCl$_3$): δ = 5.22 (dd≈t, 1H, H–3), 5.08 (dd≈t, 1H, H–4), 5.01 (dd,1H, H–2), 4.58 (d, 1H, H–1), 4.26 (dd, 1H, H–6), 4.15 (dd, 2H, H–6', OCH$_a$H$_b$), 3.83 (m$_c$, 1H, OCH$_a$H$_b$), 3.73 (m$_c$, 1H, H–5), 3.47(m, 2H, CH$_2$Br), 2.09, 2.07, 2.03, 2.01 (each s, 12H, 4 COCH$_3$); $J_{1,2}$ = 7.9, $J_{2,3}$ = 9.5, $J_{3,4}$ = 9.5, $J_{4,5}$ = 9.8, $J_{5,6}$ = 4.8, $J_{5,6'}$ = 2.6, $J_{6,6'}$ = 12.3 Hz.

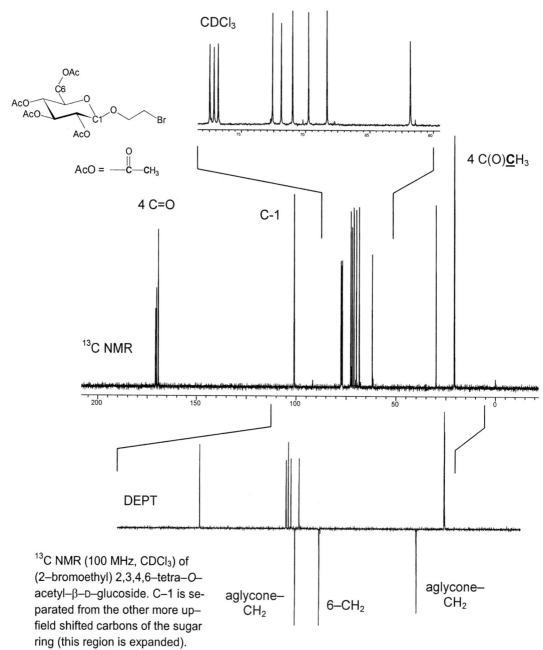

^{13}C NMR (100 MHz, CDCl$_3$) of (2–bromoethyl) 2,3,4,6–tetra–O–acetyl–β–D–glucoside. C–1 is separated from the other more up–field shifted carbons of the sugar ring (this region is expanded).

In the DEPT sequence shown, CH and CH$_3$ groups appear as positive signals, and CH$_2$ groups (C–6 and the aglycone–CH$_2$ groups) as negative signals. Spectroscopic data obtained by ^{13}C NMR are reported as follows: **(2–Bromoethyl) 2,3,4,6–tetra–O–acetyl–β–D–glucopyranoside** (for synthesis cf. chapter 4): ^{13}C NMR (62.9 MHz, CDCl$_3$): δ = 170.96, 170.57, 169.75, 169.59 (4 C=O), 101.35 (C–1), 73.05 (C–4), 72.96 (C–2), 72.27 (C–3), 68.69 (OCH$_2$), 68.11 (C–5), 62.21 (C–6), 30.30 (CH$_2$Br), 21.15 (4 COCH$_3$) ppm.

^{13}C NMR measurements are an important supplement for the NMR analysis of an unknown compound. They are, however, less sensitive than proton NMR measurements and require more material. For standard ^{13}C NMR, spectra are recorded with suppression of the C–H hetero couplings, to avoid overcrowding the spectra with peaks. While the integration information is lost in standard ^{13}C NMR because the pulse sequences applied do not allow complete ^{13}C relaxation, additional information for the interpretation of ^{13}C NMR can be obtained by so–called DEPT experiments. The information obtained in this way enables quaternary, tertiary, secondary and primary carbon atoms to be distinguished on the basis of their different ^{13}C relaxation rates. In the most frequently used DEPT sequence, quarternary carbons are suppressed, CH and CH$_3$ groups appear as positive peaks and CH$_2$ moieties as negative peaks.

To improve the resolution particularly of ^{1}H NMR spectra, the compound under investigation may be dissolved in a different solvent. Substitution of CDCl$_3$ by [D$_6$]–acetone, for example, often improves a proton spectrum significantly. Polar derivatives, which are normally recorded in [D$_4$]–methanol, may also be measured in [D$_6$]–DMSO, for example, in which the protic hydrogen atoms are not exchanged and thus also couple and resonate under the conditions of NMR. It should be noted that in general, ^{1}H NMR spectra of O–acylated carbohydrate derivatives are easier to interpret than the spectra of their O–ether protected analogs.

To determine the structure of more complicated carbohydrate derivatives, simple ^{1}H and ^{13}C NMR experiments may not be sufficient. Many sophisticated pulse sequences have been introduced, which record two–dimensional NMR spectra and can therefore provide more information than the one–dimensional spectra. These techniques facilitate the structural elucidation of very large and complicated molecules, such as complex oligosaccharides or glycoconjugates. The most common two–dimensional spectra are ^{1}H–^{1}H–COSY (<u>co</u>rellated <u>s</u>pectroscop<u>y</u>) and ^{1}H–^{13}C–COSY spectra. Examples of this kind of correlated spectra are shown for (2–bromoethyl) 2,3,4,6–tetra–O–acetyl–β–D–glucopyranoside, which also serves as a good model for those learning to interpret two–dimensional spectra.

In the ^{1}H–^{1}H–COSY the proton spectrum of a compound is correlated to itself. This leads to a symmetrical 2D–plot, in which the ^{1}H spectrum appears as projection on the diagonal of the square. So–called cross–peaks are located around this diagonal in a symmetrical manner. Thus, the part left of the diagonal contains the same information as its mirror image on the right of the diagonal. Cross–peaks indicate coupling of the multiplets in a perpendicular position to a particular cross–peak. Interpretation begins at a multiplet whose identity is known, for example the anomeric H–1. The H–1–related cross–peak leads to H–2, which then leads to the identification of H–3, H–4, and H–5. For H–5 three cross–peaks are detected, indicating its coupling to H–4, as well as to H–6 and H–6'. The cross–peak between H–6 and H–6' indicates the vicinal coupling constant $J_{6,6'}$. The aglycone protons are not in correlation with the hydrogens of the sugar ring, but are cross–peak–correlated to each other.

Similarly, multiplets for hydrogens can be cross–related to the carbon atoms to which they are attached in a ^{1}H–^{13}C–COSY spectrum. Every peak of a ^{13}C spectrum can thus be assigned, provided that the proton NMR of this compound has been solved.

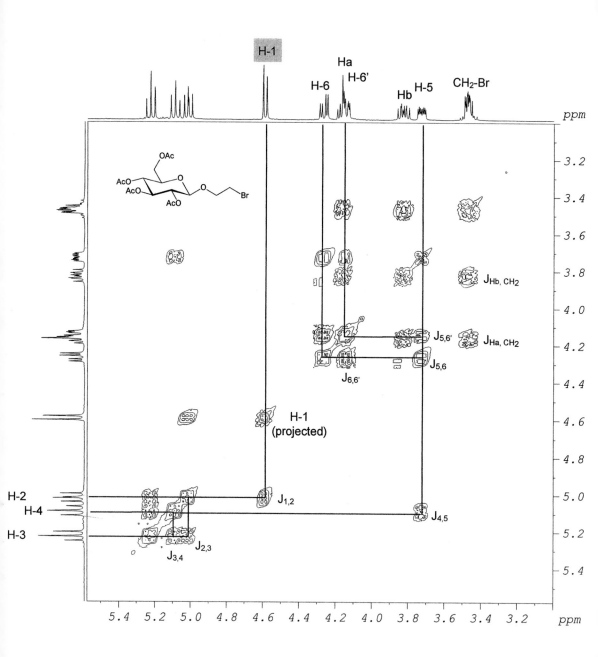

^{1}H–^{1}H–COSY of (2–bromoethyl) 2,3,4,6–tetra–O–acetyl–β–D–glucoside. Cross–peak correlations are shown by perpendicular and parallel lines.

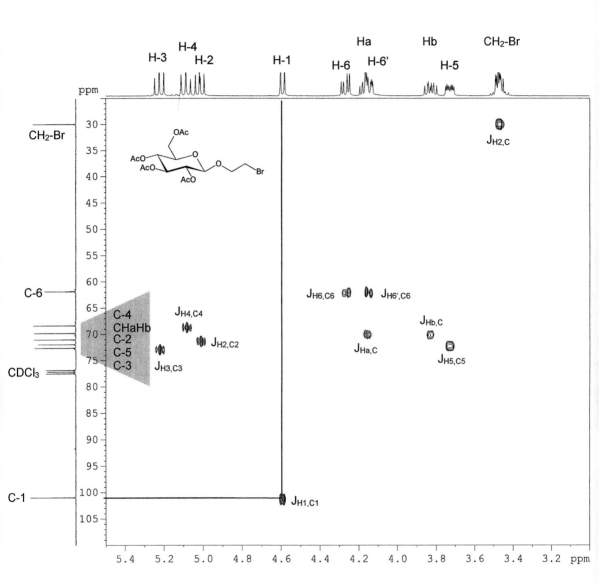

^{1}H–^{13}C–COSY of (2–bromoethyl) 2,3,4,6–tetra–O–acetyl–β–D–glucoside. All multiplets recorded in the ^{1}H spectrum of the compound can be cross–peak–correlated to the carbon atoms to which these hydrogens are attached. This is indicated for the H–1–C–1 correlation. This method facilitates the exact assignment of all ^{13}C peaks.

Mass spectroscopy

Mass spectrometric methods have become increasingly important for the structural analysis of large carbohydrate molecules such as polysaccharides and complex oligosaccharides, the structure of which cannot be unequivocally determined by NMR methods only. Moreover, mass spectrometry requires far less material than NMR spectroscopy, a fact that is of great relevance for glycobiological studies.

Fast atom bombardement (FAB), a soft ionization technique introduced in the early 1980s, was the first to revolutionize mass spectrometric studies of biopolymers. FAB mass spectrometry is used for molecular weight determination of unstable molecules, and for sequence assignments in oligosaccharides as the fragmentation behavior of oligosaccharides and glycoconjugates is well exploited [cf. A. Dell in *Adv. Carbohydr. Chem. Biochem.* 45 (1987) 19; and A. Dell and H. R. Morris in *Science* 291 (2001) 2351]. FAB-MS cannot however, be used as the sole method to determine the complete structure of a carbohydrate.

Recently, additional ionization techniques have gained importance in the mass spectroscopic analysis of glycoconjugates and neoglycoconjugates for example. These techniques include electrospray ionization (ESI) and matrix assisted laser desorption ionization (MALDI). These modern techniques also permit the ionization of molecules with molecular masses higher than 100 kDa. They are currently used extensively in biological chemistry and glycobiology in particular.

Broadly speaking FAB-MS, MALDI-MS and ES-MS can be exploited in two general ways in the glycobiology field. The can be used for the detailed characterization of purified individual neoglycopolymers and glycopolymers, or mixtures thereof. These can hardly be structurally characterized by NMR methods. Rapid screening of cell– and tissue–extracts is a second field of application as a limited number of MS experiments are often sufficient to answer the questions addressed in this case.

Each MS method has its own strengths. MALDI-MS can be the method of choice for molecular weight profiling of polysaccharide mixtures, whereas FAB-MS is ideally suited to screening for non-reducing epitopes in biological samples after appropriate treatment. Nano-ES mass spectrometry can be favorably employed in MS-MS experiments. Together with a Q-TOF analyzer, which has been introduced in the mid-1990s and has both, a quadrupole and TOF (time of flight) analyzer, nano-ES-MS-MS provides the most sensitive means of sequencing peptides and glycopeptides.

9   The literature of carbohydrate chemistry

The early studies on carbohydrates in the nineteenth and beginning of the twentieth century were published mainly in the German general chemical journals such as *Liebigs Annalen der Chemie*, often just called 'Liebigs Annalen', *Berichte der Deutschen Chemischen Gesellschaft*, often abbreviated to 'Chemische Berichte', and *Zeitschrift für physiologische Chemie*. Today, these three journals no longer exist as such since Liebigs Annalen and Chemische Berichte were incorporated into *European Journal of Organic* and *Inorganic Chemistry* respectively, in the nineties.

During the twentieth century, carbohydrate chemistry developed worldwide and reports of the findings have been published in international general organic chemical and biochemical journals and books, increasing the percentage of carbohydrate–related literature rapidly. In addition to full papers and notes in carbohydrate research, many of the general chemistry journals include review articles in their issues, and these are regularly dedicated to carbohydrates. The volumes of *Accounts in Chemical Research, Chemical Reviews* and *Reviews of the Chemical Society, Angewandte Chemie* (International Edition in English) and, more recently, *European Journal of Organic Chemistry* and *Chemistry a European Journal* are worth browsing through for reviews related to saccharides. Some of these reviews have already become classical carbohydrate literature, such as Paulsen's and Schmidt's overviews on oligosaccharide synthesis published in *Angewandte Chemie* in 1982 and 1986 respectively, in addition to Fraser–Reid's contributions entitled 'Some Progeny of 2,3–Unsaturated Sugars –They Little Resemble Grandfather Glucose' published in *Accounts of Chemical Research* in 1975 and continued in 1985 and 1996.

Moreover, a number of specialized journals and book series deal with carbohydrate chemistry and glycobiology (cf. Table 9-1). The year 1965 saw the introduction of the first specialized journal with the appearance of *Carbohydrate Research*, which was followed by a second international journal for carbohydrate research in 1982, the *Journal of Carbohydrate Chemistry*. About the same time, in 1981, *Carbohydrate Polymers* was launched and only very recently, in 1995, *Carbohydrate Letters* was added to the collection. With the development of glycobiology came the publication of even more specialized journals such as *Glycoconjugate Journal* in 1984 and *Glycobiology* in 1991. The journal *Current Opinion in Strutural Biology* includes one important issue on glycoconjugates every year.

Indispensable for the carbohydrate chemist is the series *Advances in Carbohydrate Chemistry and Biochemistry*, a compendium of detailed reviews, about five per volume, which reflect the knowledge and interests of the time. The series was started in 1945 as *Advances in Carbohydrate Chemistry* with approximately one new volume appearing every year, changing its name in 1969 to *Advances in Carbohydrate Chemistry and Biochemistry*. Modern carbohydrate nomenclature was the subject of a 1997 issue of *Advances of Carbohydrate Chemistry and Biochemistry* and was also covered in *Carbohydrate Research* and *Journal of Carbohydrate Chemistry*.

During the development of carbohydrate chemistry in the twentieth century a number of specialized books as part of a general series on organic chemistry were dedicated to carbohydrates and they remain useful references with numerous original articles cited therein. For example, part of Rodd's *Chemistry of Carbon Compounds*, volume 1F (1967) and its supplements offer useful general surveys on sugars. As part of the series *Topics in Current Chemistry* several issues have dealt with carbohydrate and glycoconjugate chemistry, such as volumes 154 (1990), 186, 187 (both 1997) and 215 (2001) and the same is true for the series *Methods in Enzymology*, which is an important source of practical information on glycochemistry and glycobiology. An impressive overview on glycosylation methods was compiled by O. Lockhoff in about 450 pages for *Houben–Weyl, Methoden der Organischen Chemie*, Band E14a/3 (1992) and included a useful survey of related review articles. In *Comprehensive Organic Synthesis*, edited by B. M. Trost, volume 6 includes a chapter on the synthesis of glycosides and one on protecting groups. In the series *Comprehensive Natural Product Chemistry*, published in 1999, volume 3 deals with the biosynthesis of carbohydrates.

An important and classical opus on carbohydrate chemistry is a four–volume work, *The Carbohydrates*, edited by W. Pigman and D. Horton in 1972. A fourth volume was added in 1980. The four volumes consist of individual chapters written by several expert authors. Another basic text in carbohydrate literature is *Methods in Carbohydrate Chemistry*, which covers detailed general procedures for the analysis and preparation of carbohydrate derivatives in nine volumes; volume I appeared in 1962, volume IX in 1995.

A long list of general and more specialized monographs on carbohydrate chemistry and glycoscience as well as on protecting groups has been published over the years and the more specialized reader is advised to consult these books and any further new editions for further study. Recently, Wiley-VCH has published four volumes on *Carbohydrates in Chemistry and Biology*, followed by Springer in 2001, where a three–volume work on *Glycoscience* has been edited. An excellent and up to date book on Glycobiology is *Essentials of Glycobiology*, edited by A. Varki et al. in 2002. Some of the older books have now become out of date, but many however, still provide accurate and fundamental descriptions of the concepts of carbohydrate chemistry and are worthwhile reading. Furthermore, the Internet has now become a useful source of information on carbohydrate chemistry.

Table 9–1. The principal texts used in carbohydrate chemistry and biochemistry.

Journals	Book series
Carbohydrate Research	*Advances in Carbohydrate Chemistry and Biochemistry*
Journal of Carbohydrate Chemistry	*Methods in Carbohydrate Chemistry*
Carbohydrate Polymers	*Methods in Enzymology*
Glycobiology	*Handbook of Oligosaccharides*
Glycoconjugate Journal	*The Carbohydrates*
Current Opinion in Structural Biology	

In addition, many reviews have been published on carbohydrate chemistry and glycobiology over the years. Some of the older ones got updated later; others never became out of date. It would take to much space to list up all the valuable work published on sugar chemistry and biochemistry, which is worthwhile reading. However, a series of review articles has recently appeared in two issues of *Chemical Reviews*, each extensively has recently appeared in two issues of *Chemical Reviews*, each extensively referencing earlier work. Therefore, the issue of *Chemical Reviews* published in December 2000, dedicated to carbohydrate chemistry, and the February issue 2002 on glycobiology are recommended for further reading. Moreover, even the journal *Science* has recently dedicated an issue (March 2001, Vol. 291) to 'Carbohydrates and Glycobiology' and this is suited to reflect the attention which is payed to the field by the entire scientific community.

List of experimental procedures

Chapter 5

Subject index